Environmental Securitisation in India and China

Joe Thomas Karackattu · Justin Joseph ·
Ramnath Reghunadhan
Editors

Environmental Securitisation in India and China

palgrave
macmillan

Editors
Joe Thomas Karackattu
Department of Humanities and Social
Sciences
Indian Institute of Technology
Madras
Chennai, Tamil Nadu, India

Justin Joseph
Centre for Climate Change
and Environment (CCCE)
Vellore Institute of Technology (VIT)
Chennai, Tamil Nadu, India

Ramnath Reghunadhan
Department of Social Sciences, School
of Social Sciences and Languages
Vellore Institute of Technology
Vellore, Tamil Nadu, India

ISBN 978-981-97-9159-0   ISBN 978-981-97-9160-6   (eBook)
https://doi.org/10.1007/978-981-97-9160-6

Cover illustration: © John Rawsterne/patternhead.com

This Palgrave Macmillan imprint is published by the registered company Springer Nature
Singapore Pte Ltd.
The registered company address is: 152 Beach Road, #21-01/04 Gateway East, Singapore
189721, Singapore

If disposing of this product, please recycle the paper.

PREFACE

The twenty-first century has witnessed an unprecedented transformation in the way environmental issues are perceived and addressed. Coupled with an escalating climate crisis characterised by rising temperatures, extreme weather events, and the degradation of natural ecosystems, this has prompted a re-evaluation of traditional security paradigms. In this era of global challenges, the interplay between knowledge, governance, and the relations between states (and non-state actors, alike) has never been more critical. Scientists and policymakers agree that humanity is approaching unfamiliar and uncharted territory, where the implications of inaction could be disastrous. The increasing consensus that we are facing a climate emergency is a result of rising global temperatures and the increasing frequency of environmental challenges. In response, international organisations have started to define climate change as a security issue, highlighting the necessity of swift and concerted action under the SDGs. This book, *Environmental Securitisation in India and China*, explores the complex interplay between environmental challenges and security concerns in India and China, two of the world's most populous and rapidly developing economies. This book seeks to illuminate the intricate dynamics that shape our understanding of global environmental policy, particularly in the context of environmental governance, security, and the evolving role of non-state actors.

The book brings together scholars, experts, and academicians to analyse environmental securitisation in India and China comprehensively. Through a multidisciplinary lens, the historical, political, and social contexts that shape diverse approaches to environmental challenges have been examined. The book is structured to facilitate a nuanced understanding of the various dimensions of environmental securitisation, including policy frameworks, institutional responses, and social changes. The contributions within this volume are a testament to the rich tapestry of thought that characterises contemporary scholarship in International Relations, which draws from a wide array of theoretical frameworks and empirical studies. With their contemporary and timeless ideas, the chapters tackle the most critical issues of our day. This book gives readers a thorough understanding of the complex subject of governance—covering everything from the effects of climate change on international security to the influence of epistemic communities on policy discourse. As we embark on this intellectual journey, we must acknowledge the diverse voices and perspectives that enrich this discourse, with the contributors to this volume hailing from various academic backgrounds, regions, and experiences. The insights serve as a reminder that no single discipline or perspective holds the key to understanding the complexities of our world. Instead, it is through collaboration and dialogue that we can forge new pathways collectively towards effective governance and sustainable solutions.

The book delves into the theoretical foundations of environmental securitisation, exploring key concepts and frameworks that inform the discourse. This grounding sets the stage for a deeper exploration of the specific cases of India and China. The subsequent sections of the book provide a detailed examination of the policies and initiatives undertaken by both countries in response to environmental challenges. The chapters highlight the role of government agencies, non-governmental organisations, and civil society in shaping environmental policy and action, exploring the impact of international agreements and collaborations on national strategies, and emphasising the importance of global cooperation in addressing transboundary environmental issues. In an increasingly interconnected world, by examining the interactions between state and non-state actors, this book highlights the importance of inclusive governance and the need for collaborative approaches to address complex global issues. The contributions also reflect a growing recognition of the role of non-state actors in shaping environmental policy, especially the involvement of environmental non-governmental organisations (ENGOs), business groups, etc.

This creates interlinkages for diverse perspectives and innovative solutions, often challenging the status quo and advocating for marginalised voices. One of the central themes that emerges from the contributions is the urgent need for a paradigm shift in approaching environmental governance. As climate change continues to pose existential threats to ecosystems and human societies, traditional governance models are proving inadequate. The book calls for reimagining the relationship with the environment, emphasising the importance of sustainability, resilience, and social justice.

This shift requires innovative policy solutions and a fundamental change in the values and priorities of a global community. In addition to addressing environmental challenges, the book also explores the intersections of security and governance. The concept of security has evolved significantly in recent decades, expanding beyond traditional military concerns to encompass a broader range of issues, including human security, environmental security, and economic stability. The book engages with these evolving definitions and discourses of security, highlighting the need for holistic approaches that recognise the interconnectedness of various threats and vulnerabilities.

As we navigate the complexities of the twenty-first century, we must recognise that the solutions to our most pressing issues often lie at the intersection of diverse disciplines, perspectives, and practices. The challenges posed by climate change and environmental security are not confined to national borders; they are global issues that require collective action and innovative solutions. As the human civilisation stands at a critical juncture in history, the lessons drawn from the experiences of

India and China can inform our collective efforts to build a more resilient and sustainable future for all. This book serves as a timely contribution to the ongoing discourse on environmental securitisation, not merely as an academic exercise but as a call to action. The book offers valuable insights, and we hope that it can serve as a valuable resource for scholars, policymakers, practitioners, and anyone interested in the intricate relationships between knowledge, governance, and International Relations.

The shortcomings remain entirely ours.

Chennai, India

Joe Thomas Karackattu
Justin Joseph
Ramnath Reghunadhan

ACKNOWLEDGEMENTS

The editors of this volume acknowledge the funding support received from the Indian Council of Social Science Research (ICSSR; F.No.02/94/2022-23/ICSSR/RP/MN/GEN), and the backend support from the Office of Industrial Consultancy and Sponsored Research (IC&SR) at the Indian Institute of Technology Madras. The editors are grateful to all resource persons that participated in the Symposium and International Conference, which resulted from this project on the broad theme of *'Greening Process' in Emerging Economies: A Comparative Survey on Public Participation in Environmental Policymaking in India and China*. We owe our sincere gratitude to Prof. Alka Acharya, Professor, Centre for East Asian Studies, School of International Studies, Jawaharlal Nehru University, New Delhi; Dr Anjan Kumar Sahu, Associate Professor, Department of Political Science, University of Allahabad, Prayagraj; Prof. Jayati Srivastava, Professor, Centre for International Politics, Organization and Disarmament, School of International Studies, Jawaharlal Nehru University; Dr Lavanya Suresh, Associate Professor, Department of Humanities and Social Sciences, BITS-Pilani, Hyderabad campus; Prof. Sreemati Chakrabarti, Chairperson, Institute of Chinese Studies (ICS) & Professor (Retd.), Department of East Asian Studies, University of Delhi, New Delhi; Prof. Varaprasad S. Dolla, Professor, Centre for East Asian Studies, School of International Studies, Jawaharlal Nehru University, New Delhi; Prof. Utham Kumar Jamadhagni, Professor & Head, Defense and Strategic Studies, University of Madras, Chennai, among others for their valuable inputs at different stages. The completion of the book project owes in no small measure to each of the contributors to the volume who were patient with us throughout the process.

Joe Thomas Karackattu
Justin Joseph
Ramnath Reghunadhan

ABOUT THIS BOOK

This book examines how emerging environmental challenges are situated within existing International Relations (IR) theoretical understandings of 'security'. Although securitisation has been studied by various scholars, environmental securitisation in Global South polities (with case studies) still remains an under-researched area despite its increasing significance in earth system governance and planetary security. As governments in the Global South undertake policy interventions to mitigate the impact of increasing climatic changes and yearn to achieve human-nature harmony, one can observe similar patterns of responses chiefly due to ecology-economy dichotomy in these states and their societies. In this context, this book brings to the readers various aspects of theory and practice of state interventions in the form of environmental securitisation in the Global South majorly under four themes (encompassing theory and policy processes). The themes explicate connections between environment and International Relations theory, securitisation in developing countries, state society and the environment in India and China and lastly, public participation and environmental policymaking. The scholarship presents a comprehensive and coherent overview on the politics of securitisation in India and China, two prominent economies in the Global South. Faculty and researchers who work on non-Western International Relations theory and non-traditional security threats, policy practitioners and experts in environmental policymaking, students of IR and Comparative Politics, chiefly, will benefit from this book.

CONTENTS

Part I   Introduction

1   Environmental Securitisation in India and China:
     An Overview   3
     Justin Joseph, Joe Thomas Karackattu,
     and Ramnath Reghunadhan

Part II   Environment and International Relations
           Theory

2   Securitisation of Development, Climate Complex
     and Global Climate Governance: The US, China
     and India   29
     Anjan Kumar Sahu

3   Problematic of Ecological Routines: Securitisation
     and Beyond   47
     Mathew A. Varghese

4   Role of Epistemic Communities in Shaping India's
     Environmental Policies: A Constructivist Perspective   59
     Chandran Komath

5   Understanding Environment and IR Theory
     from Non-western International Relations Theory:
     State-Society Interface   81
     Vikas Kumar and Ram Babu

Part III   Trends in Environmental Securitisation

6   Identity Politics and Chinese Climate Securitisation   97
     Juha A. Vuori

7   Environmental Policy in China and Interaction
     with Border Countries (the Case of Russia)   115
     Olga Zalesskaia

8   Global South and the Resistance to Pan-Securitisation
     of Environmental Issues: A Case Study of India
     and China   127
     Ashmita Rana and Justin Joseph

Part IV   State, Society and Environmental Policy
          Processes in India And China

9   Climate Securitisation in China and India: A Human
     Security Perspective   145
     Shilpi Ghosh and Gajendra

10   Tagged: The Role of Social Media in Influencing
     Environmental Governance in China and India   161
     Divisha Srivastava

11   From Policy to Practice: China's NEV Initiatives
     and the Evolution of Ecological Civilisation   179
     Akhilesh Kumar and Varaprasad S. Dolla

12   India's Experiment with Electric Vehicles: Promise
     and Ambiguity in the Context of Environmental
     Security   201
     S. Shekin and Haans J. Freddy

Index   215

Editors and Contributors

About the Editors

Dr. Joe Thomas Karackattu is an Associate Professor with the Humanities and Social Sciences Department at the Indian Institute of Technology Madras (IIT Madras). Dr Karackattu was Fox Fellow (2008–2009) at Yale University and was also the inaugural Centenary Visiting Fellow (nominated award) at SOAS, University of London, in 2013. Most recently, he was a CISLI Fellow with the India-China Institute at the New School, New York (2017–19). He serves as a Network Editor for H-Asia, part of H-Net: Humanities and Social Sciences online (based at Michigan State University) and has also served as article editor/reviewer for several Scopus-indexed journals. Dr Karackattu studied Economics at St. Stephen's College (Delhi) and Chinese studies at Jawaharlal Nehru University (Delhi). He was awarded the "Srimathi Marti Annapurna Gurunath Award for Excellence in Teaching" 2024 from IIT Madras, and was previously recipient of the "Young Faculty Recognition Award" 2020 at IIT Madras. Besides publishing research he also makes nonfiction films in order to make research findings accessible to wider audiences through the visual medium. His films have been screened at invitational events in the United States (Duke University, NYU, Penn State University, University of Pittsburgh, New School, Yale University & SUNY Binghamton), Singapore, Qatar, UAE, Denmark, France, and across India and China.

Dr. Justin Joseph currently works as Assistant Professor of School of Social Sciences and Languages (SSL) at Vellore Institute of Technology University (VIT Chennai Campus). Previously he was associated with GITAM School of Humanities and Social Sciences, Hyderabad, India. Before joining GITAM, he headed the Department of Political Science, School of Liberal Arts and Applied Sciences, Hindustan Institute of Technology and Science (HITS), Chennai, India, from its inception in May 2020-23. He obtained his PhD in Politics and International Relations from IIT Madras and was awarded the 'General Scholar Fellowship' by the Ministry of Human Resources and Development, Government of India, for conducting PhD fieldwork in China from September 2017 to July 2018. He was Visiting Scholar at Shanghai International Studies University (SISU) and Shanghai University during 2016-17 and 2017-18, respectively. He was conferred with the Pre-Doctoral Fellowship by IIT Madras for the early submission of his PhD thesis in 2020. His research areas include environmental politics in India and environmental security in the Global South.

Dr. Ramnath Reghunadhan is an Assistant Professor at School of Social Sciences and Languages (SSL) at VIT University (Vellore Campus). He did his PhD in the Department of Humanities and Social Sciences, Indian Institute of Technology Madras (IIT Madras), and works on Science, Technology and Innovation (STI) policy, development, and diplomacy. He was selected for the IIT Madras Institute Research Award 2022, the Mira Sinha-Bhattacharjea Award 2019 (by the Institute of Chinese Studies, New Delhi), and the Huayu Enrichment Scholarship programme (by the Ministry of Education, Taiwan). His book '*Cyber Technological Paradigms and Threat Landscape in India*' was published by Palgrave Macmillan (Springer). His contribution to the Government of India was selected amongst the 'Top 50' ideas for the upcoming Science Technology and Innovation Policy (STIP) by the STIP Secretariat (India) and Science Policy Forum, New Delhi. He is currently a reviewer for several Scopus-indexed journals. Moreover, he has published in journals like *International Affairs, International Studies, Round Table, Journal of Asian Security and International Affairs, Asian Affairs, Strategic Analysis, Journal of Black Studies*, and chapters published in *IoT and Analytics in Agriculture* (Springer) and *Handbook of Research on Blockchain Technology* (Elsevier).

Contributors

Ram Babu Department of Political Science, Guru Ghasidas Central University, Chhattisgarh, India

Varaprasad S. Dolla Centre for East Asian Studies, School of International Studies, Jawaharlal Nehru University, New Delhi, India

Haans J. Freddy Department of Political Science, Madras Christian College (Autonomous) (MCC), Chennai, Tamil Nadu, India

Gajendra Centre for Diaspora Studies, Central University of Gujarat, Gandhinagar, India

Shilpi Ghosh Centre for African Studies, School of International Studies (SIS), Jawaharlal Nehru University (JNU), New Delhi, India

Justin Joseph Centre for Climate Change and Environment (CCCE), Vellore Institute of Technology (VIT), Chennai Campus, Tamil Nadu, India

Joe Thomas Karackattu Department of Humanities and Social Sciences, Indian Institute of Technology Madras, Chennai, Tamil Nadu, India

Chandran Komath Department of Political Science, Government College Kottayam, Kottayam, Kerala, India

Akhilesh Kumar Centre for East Asian Studies, School of International Studies, Jawaharlal Nehru University, New Delhi, India

Vikas Kumar Department of Political Science, Guru Ghasidas Central University, Chhattisgarh, India

Ashmita Rana School of International Studies (SIS), Jawaharlal Nehru University (JNU), New Delhi, India

Ramnath Reghunadhan School of Social Sciences and Languages, Vellore Institute of Technology (VIT) University, Vellore, Tamil Nadu, India

Anjan Kumar Sahu University of Allahabad, Allahabad, India

S. Shekin Department of Political Science, Madras Christian College (Autonomous) (MCC), Chennai, Tamil Nadu, India

Divisha Srivastava South Asian University (SAU), New Delhi, India

Mathew A. Varghese School of International Relations and Politics, Mahatma Gandhi University, Kottayam, Kerala, India

Juha A. Vuori Department of Philosophy, Contemporary History, and Political Science, University of Turku, Turku, Finland

Olga Zalesskaia Faculty of Foreign Languages, Blagoveschensk State Pedagogical University, Blagoveshchensk, Russia

ABBREVIATIONS

ADB	Asian Development Bank
AMRUT	Atal Mission for Rejuvenation and Urban Transformation
APEC	Asia-Pacific Economic Cooperation
ARAI	Automotive Research Association of India
ASEAN	Association of Southeast Asian Nations
ASEZs	Advanced Social and Economic Development Zones
BIMSTEC	Bay of Bengal Initiative for Multi-Sectoral Technical and Economic Cooperation
BRI	Belt and Road Initiative
BRICS	Brazil, Russia, India, China and South Africa
CAN	Climate Action Network
CBD	Convention on Biological Diversity
CBDR	Common But Differentiated Responsibilities
CCP	Chinese Communist Party
CND	China's National Defence in New Era
CNSC	Central National Security Commission
COP	Convention of Parties
COVID	Corona Virus Disease
CPC	Communist Party of China
CSCA	Coalition for Sustainable Climate Action
CSE	Centre for Science and Environment
DPRK	Democratic People's Republic of Korea
DRC	Democratic Republic of Congo
EAS	East Asian Summit
EIA	Environmental Impact Assessments
ETF	Ecological Task Force

EU	European Union
FC	Fuel Cells
FRA	Forests Rights Act
GCF	Green Climate Fund
GDP	Gross Domestic Product
GEF	Global Environment Facility
GHG	Green House Gas
GIFT	Gujarat International Finance Tec-City
GNI	Green Nicobar Island
GRP	Gross Regional Product
HADR	Humanitarian Assistance and Disaster Relief
ICEs	Internal Combustion Engines
IDSA	Institute for Defence Studies and Analyses
IMF	International Monetary Fund
IPCC	Intergovernmental Panel on Climate Change
IR	International Relations
IRT	International Relations Theory
ISA	International Solar Alliance
IUCN	International Union for Conservation of Nature
JNNURM	Jawaharlal Nehru National Urban Renewal Mission
JST	Just Securitisation Theory
LED	Light-Emitting diode
LiFE	Lifestyle for Environment
MEA	Ministry of External Affairs
MGNREGA	Mahatma Gandhi National Rural Employment Guarantee Act
MoU	Memorandum of Understanding
MW	Mega Watt
NAPCC	National Action Plan on Climate Change
NATO	North Atlantic Treaty Organisation
NC&RE	Non-Conventional & Renewable Energy
NDF	National Disaster Relief
NDRC	National Development and Reform Commission
NEV	New Energy Vehicles
NGO	Non-Governmental Organisations
NIEO	New International Economic Order
NIH	National Institute of Hydrology
NISE	National Institute of Solar Energy
NITI Aayog	National Institution for Transforming India Aayog
NWDA	National Water Development Authority
PEMFC	Proton-Exchange Membrane Fuel Cell
PESA	Panchayats (Extension to Scheduled Areas)
PMUY	Pradhan Mantri Ujjwala Yojna
PPP	Public Private Partnerships

PRC	People's Republic of China
R&D	Research and Development
SBA	Swachh Bharat Abhiyan
SCM	Smart City Mission
SCO	Shanghai Cooperation Organisation
SDG	Sustainable Development Goal
SEPA	State Environmental Protection Administration
SEZ	Special Economic Zone
SIDS	Small Island Developing States
SOFC	Solid Oxide Fuel Cell
SPV	Special Purpose Vehicles
STAR-C	Solar Technology Application Resource Centre
TERI	The Energy and Resources Institute
UAPA	Unlawful Activities (Prevention) Act
ULB	Urban Local Bodies
UN	United Nations
UNEP	United Nations Environment Programme
UNFCCC	United Nations Framework Convention on Climate Change
UNPKOs	United Nations Peace Keeping Operations
UNSC	United Nations Security Council
USD	United States Dollar
VVT	Vehicle and Vessel Tax
WEO	World Economic Outlook
WHO	World Health Organisation
WMO	World Meteorological Organization
WWF	World Wildlife Fund

LIST OF TABLES

Table 6.1   Macrosecuritisation Elements of China's climate change
discourse   108

Table 11.1   Four Modernisation Theories & Their linkages
with NEV Policies   194

Table 11.2   Here is a table about how NEV Policies are linked
with SDG Goals in China   195

Introduction

Environmental Securitisation in India and China: An Overview

Justin Joseph⦿*, Joe Thomas Karackattu, and Ramnath Reghunadhan*

Abstract Securitisation framework of the Copenhagen School in International Relations Theory has widely been applied to study policy processes aimed at environmental security, however, in the developed liberal democracies of the Global North. Non-Western polities such as India and China also attempt to securitise their environmental sector due to emerging

J. Joseph (✉)
Centre for Climate Change and Environment (CCCE), Vellore Institute of Technology (VIT), Chennai Campus, Chennai, Tamil Nadu 600127, India
e-mail: justinj8064@gmail.com

J. T. Karackattu
Department of Humanities and Social Sciences, Indian Institute of Technology Madras, Chennai, Tamil Nadu, India
e-mail: joe@iitm.ac.in

R. Reghunadhan
School of Social Sciences and Languages, Vellore Institute of Technology (VIT) University, Vellore, Tamil Nadu, India
e-mail: vrramnath@gmail.com

J. T. Karackattu et al. (eds.), *Environmental Securitisation in India and China*, https://doi.org/10.1007/978-981-97-9160-6_1

non-traditional security challenges like pollution and natural calamities. One of the significant differences between countries of the Global North and Global South in this context is that the latter witnesses a policy dichotomy due to their economic interests and ecological conditions as development is understood as the anti-thesis of environmental conservation. This chapter sets the ground for exploring the patterns of differences and similarities in the process of environmental securitisation in India and China for gathering more evidences from the Global South for theory building in International Relations in general and the role of audience in securitisation framework in particular.

Keywords Securitisation · Audience · Global South · India · China

Introduction

Environmental policy processes in the Global South are increasingly complex (Aryal et al., 2021) due to various factors such as economy–ecology dichotomy, transnationality of socio-environmental challenges and involvement of non-state interest groups. With the end of the Cold War, there is a significant proliferation of non-state actors in world politics (Krahmann, 2005). They continue to influence local, regional and national policy processes (Weiss et al., 2013) especially during the pursuit of economic development. These developments have, arguably, widened the dialogic space between the individual and the state, and consequently new structures of interaction have emerged which have also given rise to new forms of deliberative-communicative mechanisms (Joseph, 2020). Economic development is undoubtedly an important policy objective for governments at various levels which then directly influences human-nature harmony. The impact of global environmental crises on states and their societies is increasingly visible in recent times. States' capacity is considered an important factor in grappling with the global environmental crisis, without trivialising the contributions of actors in the non-state domain (Brauch, 2008).

States' relations with other counterparts and their relationship with their respective societies take place in the context of increasing global interconnectedness and global environmental crises. As a result, the 'environment' as a sector is capable of inducing changes in current forms of state-society interactions with the possible emergence of an environmental security discourse. According to Burchill et al. (2022), "the political force of social and intellectual movements centred on environmentalism (and also feminism), have undermined the safe and traditional certainties of International Relations' thought: as a result, the intellectual boundaries of International Relations have been contested and are being redrawn" (Burchill et al., 2022). Newly emerging systemic challenges from international politics in general and global environmental crises in particular increase the complexity of interaction among the states as well as extrapolate the theoretical borders of IR theories to encompass emerging norms in the post-Cold War world order.

This "normative competition and complexity" needs more attention as "world politics turns increasingly multipolar, as well as post- Western" (Acharya, 2017; Acharya &Buzan, 2009; Tickner & Wæver, 2009). A more comprehensive and genuinely interactive dialogue would benefit both, the Western and non-Western IR scholarship (Lu et al., 2024). This situation presents us two important tasks, one is to look for evidences for theory building outside Western discourses and secondly to explore patters of similarities and dissimilarities in the policy processes in the Global South countries. The objectives of this volume are twofold—first, to gather theoretical evidences in the application of securitisation theory outside of the 'Western' settings in order to contribute to the strengthening of IR Scholarship emerging from this specific context. The book also aims to produce practical accounts on the role of audiences in the securitisation processes in non-Western developing economies such as India and China.

Conceptualising 'Greening' as a Process Initiated by State and Undertaken by Society

Following a social constructivist approach to examine environmental policymaking and its transformation towards human-nature harmony, the volume mirrors the pursuits of other scholars such as Acharya and Buzan (2009) to "challenge traditional IR theories to accept the ideas, experiences, and insights from the non-Western worlds, and expect them to give due recognition to the place, roles and contributions of non-Western people and societies". For Constructivists, normative and ideational structures are of significant importance because institutional meanings are vital in constructing the social identity of actors. In connection with this, the interests and actions of the actors are constituted by that identity (Wendt, 1995). Hence agents and structures are equally important in Constructivism, and much of the Constructivist literature examines the reciprocal interaction between agency and structure, unlike the traditional state-centric theories of International Relations. In this context, greening is understood as a constructive process by the state and its agencies to escape the stemming global environmental crisis. Greening can be defined as the process of inducing an ecological turn at the policy levels. Constructivist frameworks help us to find the most suitable theory to examine this process of greening precisely because of the inter-subjective nature of policy processes, i.e. state action and societal response. Considering the state as a significant actor in the environmental sector in Global South, the book draws insights from the literature on the Copenhagen School to examine the state actions and policy processes in the environmental sector in two significant growing economies with highest population in the world, namely India and China. As the Copenhagen School goes global to unravel the securitisation in the non-Western polities (Kapur & Mabon, 2018), the book examines environmental securitisation in India and China to discover the underlying changes in state action and the shifts in the nature of agent-structure dynamic in the context of vertical deepening and horizontal widening of the traditional concept of security (Buzan et al., 1998).

The Copenhagen school differs from conventional Security Complex Theory by taking "an explicitly social constructivist approach to understanding the process by which issues/sectors become securitised" (Buzan et al., 1998). Environmental sector can be securitised similar to the military sector and can be elevated to high political priorities through state interventions. The overall political process of this elevation is known as 'securitisation'. According to Waever (2008), "securitisation is the discursive political process, through which an inter-subjective understanding is constructed within a political community to treat something as an existential threat to a valued referent object and to enable a call for urgent and exceptional measures to deal with the threat" (Waever, 2008). When there is an existential threat on humanity because of global environmental crises, securitisation of the environmental sector legitimises unprecedented actions to address this threat. This includes extraordinary measures such as demarcating Ecological Sensitive Areas (ESA) as seen in the case of India's Western Ghats conservation (Gadgil, et. al., 2011) or deploying environmental police as seen in the case of China (Van der Kamp, 2021).

The state (or the political authority) is inseparable from securitisation process as it is the only legitimate entity to adopt extraordinary measure to mitigate the existing challenge. In the context of increasing demands for human-nature harmony, national governments adopt policies towards conserving their natural resources as part of securitising the environment. Hence environmental securitisation can be defined as policy interventions from the state to ameliorate 'threats from human activity to the natural systems or structures of the planet when the changes made do seem to pose an existential threat to civilisation' (Buzan et al., 1998). The post-2020 global biodiversity framework (2021) identified human-nature harmony as constructive and achievable through political will at the highest level of government (CBD, 2021). Such political will can effectively be studied through securitisation framework, which argues that a series of political processes such as the role of political executive through continuous performative speech acts, declaration of extraordinary measures, acceptance by the audience and success of extraordinary measures can securitise things. As it is an inter-subjective exercise, the process of securitisation involves seven major factors (Buzan et al., 1998; Vuori, 2024). According to Vuori (2024):

1. A "securitising actor" ("that which or who makes the move towards a new or to alter an existing issue of security in accordance with particular conventions and grammars");
2. A "referent object (that which is to be secured)";
3. A "threat" (that which "threatens the referent object)";
4. An "audience" ("the necessary relation needed to produce the deontic modality of security or those who have to be 'convinced' for securitisation to be satisfied");
5. "Felicity conditions" ("rules and conventions of the speech act and its consequences");
6. "Facilitation factors" ("factors that can facilitate or impede the acceptance of the securitisation move"; "social conditions that relate to social positions of the actor and audience as well as the threat");
7. "Functional actors" ("actors that are neither the securitising actor, the threat, nor the referent object but still have some bearing on the process)".

It is also worth noting that these components of securitisation vary according to changing nature of the political systems. One can observe similarities and differences among these factors in the processes of environmental securitisation in India and China. As mentioned earlier, securitising actor is always the political authority in both the cases. However, the contexts in which these political authorities operate and their behaviour differ primarily due to the differences in their respective political systems. Consider for instance, though the referent object during environmental securitisation in China prima facia would be human-nature harmony. However, it is practically the legitimacy of the Communist Party of China (CPC) that is under threat due to the emerging environmental challenges in the country. Hence, consistent economic growth and social stability are important for the CPC to enjoy legitimacy among the citizens. In the case of environmental securitisation in India, which is a liberal democratic political system, the referent object is more accountable to audience and it is human-nature harmony at local, regional and national levels. In the case of policy interventions in India, the referent object would be human-nature harmony, securitising actor would be the Indian state, and the state is trying to convince the audience about the necessity of extraordinary measures.

The audience in environmental securitisation in both the Indian and Chinese cases are the respective citizenry. Now, extraordinary measures in securitisation should also be understood as a relative concept with respect to the history of particular polities. Consider for instance for environmental security, extraordinary measures in the case of Global North could bypass public opinion and legislative procedures (Floyd, 2016) and in the case of non-democratic polities like China it can be offering avenues for public opinion (Joseph & Karackattu, 2022). Public participation is not an extraordinary measure in democracies, but if there are law reforms that guarantee public participation in environmental policy processes, it is an extraordinary measure in the Chinese political context. Functional actors include various stakeholders and non-state actors such as Environmental Non-Governmental Organizations (ENGOs) and Business enterprises. This is the same in the Indian context and functions actors are more powerful and legitimate. Felicitation factors include the conditions that influence the process of environmental securitisation in the polities. Though there are visible dissimilarities between facilitation factors in India and China, one can also observe patterns of similarities as well. The penetration of Internet among the audience and the channels of communication it offers to participate in the policy processes through new media activism is an important factor that affects environmental securitisation in these countries. Similarly, the growth of renewable energy sectors that supports securitisation of the environment without adversely affecting economic growth is also an emerging trend that supports environmental securitisation in India and China. As sources of environmental issues are predominantly industrial and economic sectors, political authorities target these sectors for greening. Greening of the industrial/economic sector is a complex process in China due to the preoccupation with economic productivity at the local, regional and national levels. Therefore, securitisation may involve tensions between local/regional economic interests and the top-down securitisation process.

In both cases of environmental securitisation in India and China, *speech acts* of the political authority play a vital role in constructing the threat discourse and gaining legitimacy to undertake extraordinary measures to deal with the threat. Extraordinary measures in the case of India's environmental security can include actions such as introducing new tools or agencies to balance human-nature harmony, changing the conventional practice of environmental conservation, adopting new policies that challenge existing state-society relations (Executive) and framing new law or acts that can redefine the roles for traditional actors (Legislative). securitisation involves less public participation than ordinary conservation as it is more about the role of the state, i.e. *speech acts* by the political authority and the extraordinary measures undertaken by them. In other words, securitisation revolves mainly around the *securitising actor* (the political leadership/ the state), *the referent object* (human security) and *functional actors* (the agencies that influence the top-down securitisation efforts from the state).

Contextualising Economy–Ecology Dichotomy in India and China

Though both India and China assumed their status as independent nation-states around the middle of the twentieth century, China initiated the "reform and opening up policy (改革开放 or *gaige kaifang*)" in 1978 and, over the next three decades, an infrastructure-intensive, export-oriented economy led to spectacular socioeconomic growth leading millions of Chinese people out of poverty (Ang, 2018). However, this has also given rise to, on the one hand, unprecedented levels of energy consumption and, on the other hand, severe environmental pollution (Kahn & Zheng, 2016; Shapiro, 2001; Song & Woo, 2008). The increase in energy usage was heavily dependent on non-renewable energy sources such as coal and petroleum products (Wang, 2014). According to Michal Meidan (2022) of Oxford Institute for Energy Studies, coal consumption accounts for most of the total energy consumption in China, and the economy depends heavily on it, making it difficult for environmental regulations (Meidan, 2022). As the ecological problems are mounting, China's political leadership faces a key challenge: to clean the air, water and soil in order to contain the fuelling discontent among the country's populace, especially the burgeoning middle class.

In the case of India, before economic liberalisation was rolled out in the 1990s as a result of the 'Washington Consensus', the country witnessed environmental movements in the 1970s in parallel to global initiatives such as the UN summit in Stockholm on Human Environment in 1972 (Shah, 2004). Chipko and Silent Valley movements in the 1970s were two successful protests in the North and South of India which witnessed the involvement of citizens against government projects (Guha, 2014; Nabhi, 2006). Since the 1990s, a large number of protests and environmental movements have been taking place in different parts of the country, chiefly concentrating on hazardous industrial projects (Nayak, 2015) accompanied by reforms in legislation such as environmental Public Interest Litigations (Sahu, 2008). However, even when the movements are increasing, India lacks comprehensive frameworks such as ecological civilisation in dealing with stemming environmental problems in the country. Environmental movements, in general, depend on the constitutional provisions to raise their concerns with the legislative, executive, and judicial branches of the government. Among them, the judiciary plays a proactive role in environmental jurisprudence when compared to the legislature and executive organs of the Indian government (Sivaramakrishnan, 2011). These movements also paved the way for the establishment of Environmental Non-Governmental Organizations (ENGOs) across the country. Unlike the Chinese context, Indian ENGOs like the Centre for Environment and Science (CSE) have longer history since 1980, because there is no legal opposition for their existence in the Indian political system.

It is also interesting to note that, despite the single-party political system, Chinese polity has a robust presence of environmental NGOs in the local, regional and national levels (Xu & Byrne, 2021). In the context of increasing environmental problems and absence of proactive government measures, ENGOs such as Friends of Nature (FON) were founded by Liang Congji (then Peking University professor and CPPCC member in Beijing in 1994). In 1999, FON had 553 individual members (Ho, 2001). In 2018, FON membership had nearly 30,000 (FON, 2018). The non-governmental organisations rose from 5700 to 7041 between 2007 and 2012 (Liu, 2013). Being ruled by a single party, the state has placed strict control on the speed of non-governmental organisations like Environmental Non-Governmental Organizations (ENGOs) through the Regulation on the Registration and Administration of Social Organization (1998). Despite the party-state controls, a large number of Environmental Non-Governmental Organisations exist in China without formally registering with the Ministry of Civil Affairs (Liu et al., 2017). Consistent growth of environmental NGOs in India and China despite the differences in political system signifies the relevance of the functions they carry out and recognition of the same by political authority. The emergence of environmental NGOs in the Chinese polity is not due to the absence of regulatory measures—instead it owes to the political authority considering ENGOs as functional actors in the process of securitising the environmental sector in China (Joseph, 2023). Besides ENGOs, other newer developments in the environmental sphere, such as the spread of 'new media' (social media, Chat handles, Blogs), began to play a significant role in China in promoting environmental public participation (Chen, 2014). Today, a keyword search for 'environmental pollution in China' in Baidu (one of China's leading search engines) would generate more than 900,000 results. With rapid rate of Internet penetration among the audience, social media users in India reached 448 million as per statistics in January 2021 and in China this comes to 999.95 million in 2021 (Varghese, 2024).

EVOLUTION OF THE POLICY PARADIGMS AS PART OF SECURITISATION PROCESS

From 'Attack on Nature' and 'Negligence of Nature' to the Securitisation of Environmental Sector in China

A discursive analysis of the politics (including those of policies, legislations and processes) of ecology in the PRC reveals the adaptability of the party-state dual governance structure in transforming itself from being a proponent of 'attack on nature' towards becoming the agent of securitisation of the environmental sector. Environmental sector during the Mao era was a source of economic development. International isolation and lack of industrial sector forced political administration to exploit natural resources for food and agriculture. Mao era witnessed series of attack on nature through policies like 'Man must conquer nature', 'Four pets' campaign', etc., during the Great Leap Forward from 1958 to 1962. Cultural Revolution during 1966–1976 also impacted natural environment in various ways during the campaigns like 'Down to the Countryside' (Economy, 2011; Shapiro, 2001). These direct attacks on nature diminished when Deng era reformed the economy since the late 1970s. Environment was not a concern for political administration as economic industrialisation was the priority (Economy, 2011; Jahiel, 1997; Sanders, 1999). However, with the politicisation of the environment during the Hu-Wen Administration and later securitisation of the environment during the Xi-Li Administration, political leaderships have introduced a new discourse on humannature harmony, particularly through state-led campaigns such as 'War on Pollution' and the Global Security Initiative (GSI).

Responding to the challenges to the party's legitimacy from these unintended consequences of rapid economic growth, one can observe a change in government policies since the mid of the 2000s in China. The leadership of Hu Jintao and Wen Jiabao introduced both the concept of 'Harmonious Society' (和谐社会, Héxié shèhuì.) and 'Scientific Outlook on Development' (学发展观, Kēxu. fāzhǎn guān) in 2005. Their vision of harmonious society and scientific development concept represents a break from the development paradigm of the previous leadership (Chan, 2010) because the Hu-Wen leadership introduced the first-ever 'renewable energy law' and 'Green GDP' in China immediately after the above-mentioned course correction strategies and "China's accession to the World Trade Organization in 2001". This was followed by the introduction of the "Environmental Impact Assessment (EIA) Law" and "Law of Circular Economy" in 2008. In the report to "18th National Congress of the CPC" on 8 November 2012, Hu Jintao observed: "faced with increasing resource constraints, severe environmental pollution, and a deteriorating ecosystem, we must raise our ecological awareness and enhance system building to promote ecological progress" (CPC, 2012). These efforts to balance economic growth and ecological progress were followed by the leadership of Xi Jinping and Li Keqiang in a more comprehensive manner (Fu & Distelhorst, 2018). The latest form of the economy–ecology outlook of the new leadership is centred on the notion of "Ecological Civilisation (生态文明 or shengtai wenming)".

'Ecological civilisation', originally highlighted by Hu Jintao in 2007, has been elevated to a constitutional principle of the PRC after the 19th national congress of the CPC (CPC, 2017). It is a top-down attempt to integrate ecological norms into "economic, political, social, and cultural" spheres of the country (Pan, 2016). Zhu Guangyao defines ecological civilisation as "ethical morality and ideology which realises harmonious co-existence and sustainable development both among people and between them and nature and society, reflecting the progress of civilisation. Introducing this concept enriches and deepens the theory of sustainable development, producing bold innovations and practices to evolve civilisation to a higher level" (Guangyao, 2016).

As a result of the coordinated efforts from the government agencies and private sectors, China became the worlds' largest renewable energy producer. According to the China Renewable Energy Outlook 2017, one of the reasons for this achievement is the integration of "ecological civilisation" to the country's overall development plan and the government's comprehensive strategies of "economic, political, cultural, and social progress" (Centre for Renewable Energy Development, 2017). China aims to "increase the renewable energy capacity by 38% in 2020 compared to 2015 levels, equalling 680 Gigawatts of installed capacities and investments of USD 361 billion in renewable energies" (UNFCCC, 2017). Along with the shift to renewables, it is also expected that China's coal production sector will have standardised coal-quality management (Xiangfei et al., 2018). According to the latest reports by Forbes, the country was also responsible for 38% of total global clean tech spending in 2023, investing an impressive $676 billion (Mendiluce, 2024).

Jonna Nyman and Jinghan Zeng (2016) find that securitisation of the energy sector is a policy response from the party-state to address emerging legitimacy crises due to increasing environmental incidents (Nyman & Zeng, 2016). They also stress for extended research on the implications of 'securitisation' on policymaking processes to encompass emerging trends at the societal and individual levels. Crucial aspects of exploration of the securitisation process in relation to the environmental sector in China include dynamics at local and regional levels due to conflict of interest between different layers of power, emerging tensions due to the interaction of 'stability maintenance (维护稳定 weihu wending)' mechanisms implemented by the party and constitutional guarantee by government apparatuses and the contributions of non-state actors. Considering the historic obsession of the party to maintain hegemony over state and society, role of non-state actors raises important questions in the context of securitisation of the environmental sector in China.

Politicisation of Environmental Issues in India

In contrast to the Chinese case, the Indian context lacks national-level extraordinary measures and policy-related macro course corrections in the context of environmental securitisation. The Nehruvian era during the initial decades of Indian republic faced a similar situation of import substitution, agricultural and industrial modernisation using natural resources similar to China. Huge hydroelectric projects were set up as 'temples' of Modern India with central government playing a significant role in modernising the country. Indian state is party to several international agreements since the 1972 United Nations Conference on Human Environment. There were mixed responses to environmental conservation from the respective governments in the country. Though the Department of Environment was established in 1980, there was no comprehensive law on environmental conservation. It was only after the industrial disaster Bhopal Gas Tragedy in 1984 that the other side of blind economic growth was revealed (Sriramachari, 2004).

Independent India formulated its first national law, Environmental Protection Act in 1986. India's participation in international summits on environment continued. Economic reforms in the 1990s brought in foreign capital and investors marking a new phase in the economy–ecology dichotomy in the country. As a policy solution to emerging pollution and environmental issues, National Environmental Policy was introduced in 2006 that intended to mainstream environmental concerns in all development activities. Subsequently, Environmental Impact Assessment Law introduced in 2006 and amended in 2020 opened up public to participate in policy processes with decentralised structures such as State Level Environment Impact Assessment Authority and State or Union Territory or District Level Expert Appraisal Committees. In addition to the EIA, Public Interest Litigation (PIL) also plays a significant role in ensuring the role of public. One of the important points of difference between Chinese political systems is that Indian judicial institutions play a proactive role in environmental PIL in the country. Alongside the policy and legal reforms, the country witnesses large number of environmental social movements for protection human-nature harmony.

In addition to changes in party-state actions, the environmental sector witnesses two other vital phenomena for greening: increasing advocacy from non-state actors and expanding new media activism due to the increasing rate of Internet penetration in China. The contributions of non-state actors such as ENGOs, private business enterprises and the new media activism, to a large extent, are legitimised by the positive changes in party-state circles and the legislative guarantee of public participation for greening the society. It is worth noting the visible transition in party-state circles, which can pave the way for deliberative engagement of non-state actors with the party-state structure. One of the significant reasons for the germination and existence of non-state actors such as ENGOs or private business organisations in Chinese polity is due to the evolution of norms, practices and institutions of party-state dual governance, signalling the "transformation of the party-state from a mere economic growth ensuring entity to" a coherent negotiator between different socioecological and politico-economic structures. Supremacy of the party is maintained in this transformation process, and it also has the features of crisis scanning procedures and policy responses towards greening (Young et al., 2015).

Along with new policy introductions such as ecological civilisation in the context of new security thinking, newly introduced 'Environmental Protection Law of the People's Republic of China' in 2015 ensures 'citizens' right to information and to participate in environmental policy-making. Participation in environmental impacts of construction projects and environmental PIL are important aspects mentioned in the Environmental Protection Law. When these laws at national levels, along with ecological civilisation, aim to build a sustainable way of life by securitising environmental sector, audience is also encouraged to shift to renewables along with state-led green technological innovation (Xu & Mao, 2022). This model of sustainability according to Xi Jinping's thought on ecological civilisation envisages the creation of a *community of shared future for humanity*. This idea of 'commonness' in environmental concerns is an important turning point in the history of environmental policymaking of countries. Recognising the significance of this, the United Nations added the idea of 'shared future for mankind' to its lexicon on environmental discourses (Zhao, 2022).

When compared to the associational role of environmental NGOs in Chinese political context, Indian ENGOs are found to have a confrontational relationship with the governments from time to time. According to a report of Ministry of Home Affairs, environmental NGO activism stalls developmental projects causing 2–3% decrease in the annual GDP growth (MHA, 2014). Despite this official view, Indian ENGOs are widespread and they play a vital role in protecting the natural ecosystem, educating the masses, bringing the cases into the public areas and instrumental in laws like Forest Rights Acts (FRA) 2006. ENGOs along with other stakeholders actively engage with state and society during environmental movements. Environmental movements today are capable of reflecting the ecological concerns of traditionally marginalised sections like the Dalits in India (Sharma, 2024).

Technology-Enabled Ecosystems of Activism in the Realm of Environment

Guobin Yang (2003a, 2003b, 2005a, 2005b, 2011) examines in detail the role of the Internet in felicitating public participation and finds that it supports the advocacy and activism of non-governmental organisations. According to Yangzi Sima (2011), Internet and ENGO advocacy emerge as a co-evolutionary development process, and this is increasingly interdependent as they support each other (Sima, 2011) in the process of securitisation of environment. While considering the germination and spread of environmental NGOs in particular and on-going securitisation in general, an account on the Internet and new media advocacy/activism would be an inevitable. The rise of new media, which transformed the 'one-to-many model' of traditional mass communication processes with the possibility of 'many-to-many' web of communication, is a dominant channel not only for ENGOs (Croteau & Hoynes, 2013). The government also uses these channels to educate and communicate with citizen (Joseph, 2020). It has allowed citizenry to express their views and opinions on different subjects through microblogs, Bulletin BillBoards, websites, videos and other user-generated media, creating virtual online communities and platforms for public discussion.

One of the areas in which non-state actions (through media activism and environmental advocacy) and state behavioural change (policy course correction) are increasingly consolidated is environmental justice movements. Displacement of environmental harm and the response of audience have now become sources of environmental justice movements based on Not in My Back Yard (NIMBY) demands (Zhang & Barr, 2013). These movements are successful largely due to the legal and policy changes as part of environmental securitisation (Joseph & Karackattu, 2023). Environmental protests against polluting factories, firms, and State-Owned Enterprises (SOEs) are increasing in China, which can alter economic outlook from non-renewables to green growth and renewable energy (Chen, 2014). One of the examples that proves bottom-up demand for environmental security and green growth is the Xiamen Movement of 2007. The protests successfully mobilised citizens by disseminating information on the harmful effects of paraxylene on the natural environment and human health (Yu & Zeng, 2010). There are comparative studies on Xiamen anti-paraxylene protests in China and Kodaikanal Mercury Poisoning in India which opine that there are similarities in the construction of environmental security narratives despite the differences in political systems (Joseph & Karackattu, 2023).

The 'greening' processes in India and China through environmental securitisation produce interesting patterns of state actions and non-state agential activities. Some of the important aspects in these patterns are distinct from securitisation in developed Western settings. While the greening process in India began in the 1970s by the non-state actors, it is largely led by the state in China since the 1990s. Environmental securitisation in China has multiple objectives beyond human-nature harmony. It aims to ensure consistent economic growth through green technological innovation and also to legitimise the political authority of the Communist Party of China. As a result, environmental movements in Chinese political system are not organic as one can observe in the context of India. They are regulated and managed by the party-state stability maintenance apparatuses so that they do not scale up to other regions. It can be observed that the nature of interactions of non-state actors with the political authority in China is largely collaborative. They exhibit functional alignment as they are considered functional actors in the process of securitising the environmental sector in the country. However, Indian non-state actors such as ENGOs confront the political authority on streets with force often leading to a confrontational relation with them. State actions towards environmental securitisation in China include extraordinary measures such as reforms in legal and regulatory mechanism to promote public participation in environmental policy process. When compared to a democratic polity like India, public participation, even though permissible, is an extraordinary measure adopted by China which is a single-party political system. Environmental securitisation in India is yet to witness a macrolevel state action that can be designated as successful extraordinary measure. In other words, environmental securitisation in India is at the politicisation phase but with a very active audience as participants. Similarly, environmental securitisation in Chinese polity also offers room for audience as participants. Contrary to the role of public as audience in securitisation in the Western settings, role of audience as participants in the processes of securitisation is a unique feature in developing economies like India and China in Global South.

Section I of the book includes four chapters that explore various aspects of theoretical insights from environmental securitisation in non-Western polities. Based on a critical examination of the existing understanding on securitisation, **Anjan Kumar Sahu** explores securitisation in India as a process exhibiting both normative and non-normative practices. As India presents a normative stance internationally (advocating for the poor domestically), it also enforces non-normative practices that perpetuate inequality, employ illegal means and militarise policies, which Anjan Kumar Sahu notes as being contrary to the popular Western understanding that normative and non-normative dimensions are integral to securitisation. By examining exceptional cases like Special Economic Zones (SEZs) and concepts like compensation in the process of greening in India, **Mathew A Varghese** identifies situations when state approaches to ecology become problematic in his chapter. The chapter also considers whether ethnographies of evolving ecologies can provide alternative perspectives for understanding and theorising securitisation dynamics from non-Western countries like India. **Chandran Komath's** chapter on Constructivism and securitisation contextualises the role of epistemic communities in shaping climate policies. Section I on Environment and International Relations Theory comprehensively lays out the role of environment to contribute to theory building from non-Western polities.

Section II of the book discusses major trends in environmental securitisation with analyses situated in 'Global South' settings. The section contains three chapters that delve into the dynamics of environmental securitisation in domestic and international milieus. **Zalesskaia Olga** examines the nuances of bilateral environmental relations between China and Russia in the context of politicisation of environmental issues in both the countries. With the support of case study evidence on the transboundary River Amur, she finds that environmental cooperation takes prominent positions in their bilateral relations to prevent environmental incidents. In their chapter on various aspects of Global South's resistance to international environmental securitisation, **Asmita Rana and Justin Joseph** highlight the Eurocentrism in international climate agenda setting. The chapter explains in detail the unique realities in Global South that resists pan-securitisation of environmental concerns with a comparative analysis of the contexts in India and China.

Section III of the book examines state actions and societal activities in the 'greening' process in India and China. **Shilpi Ghosh and Gajendra** compare the changing security dynamics in India and China by cataloguing policy interventions in the context of changing security discourses in these countries (supported also by the audience as participants). In the same section, **Divisha Srivastava** takes the focus on audiences in depth by examining the use of channels such as new media to communicate with the securitising actor, resulting in interesting dynamics of greater audience participation in environmental securitisation discourses. Green initiatives such as green technological innovations in the changing context of security discourses in India and China are the focal point of discussion in **Akhilesh Kumar and Varaprasad S. Dolla's chapter which brings out insights on** how environmental securitisation is aimed at mitigating pollution and energy-efficiency issues in the automobile sector in China. A similar vantage point on green innovation as part of environmental securitisation in India is offered in the chapter by **Shekin Sancho and Haans J Freddy** on government initiatives and audience responses in the Electric Vehicle sector.

The new political realities in the Global South are opportunities to reimagine the empirical and theoretical boundaries of existing IR theories. The role of societal factors is not sufficiently addressed in the existing securitisation framework. The application of the securitisation framework can reveal processes like the emergence of non-state agencies in polities such as China, besides bringing out other manifestations of public participation in environmental policymaking in order to collectively reimagine our understanding on actors, processes and outcomes in the construction of the securitisation framework itself. Hopefully this volume will contribute meaningfully to the debate by bringing in established academics and emerging scholars with expertise in this area.

REFERENCES

Acharya, A. (2017). Theorising the international relations of Asia: Necessity or indulgence? Some reflections. *The Pacific Review, 30*(6), 816–828. https://doi.org/10.1080/09512748.2017.1318163

Acharya, A., & Buzan, B. (2009). Why is there no non-Western international relations theory? An introduction. *Non-Western international relations theory* (pp. 11–35). Routledge.

Ang, Y. Y. (2018). *How China escaped the poverty trap.* Cornell University Press.

Aryal, K., Laudari, H. K., Neupane, P. R., & Maraseni, T. (2021). Who shapes the environmental policy in the global south? Unpacking the reality of Nepal. *Environmental Science & Policy, 121*, 78–88. https://doi.org/10.1016/j.envsci.2021.04.008

Bai, X., Ding, H., Lian, J., Ma, D., Yang, X., Sun, N., Xue, W., & Chang, Y. (2018). Coal production in China: Past, present, and future projections. *International Geology Review, 60*(5/6), 535–547. https://doi.org/10.1080/00206814.2017.1301226

Benney, J. (2016). Weiwen at the grassroots: China's stability maintenance apparatus as a means of conflict resolution. *Journal of Contemporary China, 25*(99), 389–405. https://doi.org/10.1080/10670564.2015.1104876

Biermann, F. (2022). The future of environmental policy in the Anthropocene: Time for a paradigm shift. *Trajectories in environmental politics* (pp. 58–77). Routledge.

Brauch, H. G. (2008). Introduction: Globalisation and environmental challenges: Reconceptualizing security in the 21st century. In H. Brauch et al. (Eds.), *Globalisation and environmental challenges: Reconceptualizing security in the 21st century. Hexagon series on human and environmental security and peace book series* (HSHES, Vol. 3, pp. 27–43). Springer.

Burchill, S., Linklater, A., Donnelly, J., Nardin, T., Paterson, M., Reus-Smit, C., ..., & Sajed, A. (2022). *Theories of international relations.* Bloomsbury Publishing.

Buzan, B., Wæver, O., & De Wilde, J. (1998). *Security: A new framework for analysis.* Lynne Rienner Publishers.

Centre for Renewable Energy Development. (2017). *China renewable energy outlook 2017 – Executive summary.* China National Renewable Energy Centre. Retrieved from http://www.cnrec.org.cn/english/publication/2017-10-18-532.html

Chan, K. M. (2010). Harmonious society. In H. K. Anheier & S. Toepler (Eds.), *International encyclopedia of civil society* (pp. 821–825). Springer.

Chen, S. (2014). *Environmental communication with Chinese characteristics: Crises, conflicts, and prospects* (Doctoral dissertation, Communication, Art & Technology), Simon Fraser University (SFU). Retrieved from http://summit.sfu.ca/item/14799

Communist Party of China (CPC). (2012). *Report to the 18th National Party Congress*. Shanghai Library.

Communist Party of China (CPC). (2017). *Report to the 19th National Party Congress*. Shanghai Library.

Convention on Biodiversity. CBD-UN. (2021). First draft of the post-2020 Global Biodiversity Framework, CBD. https://www.cbd.int/doc/c/abb5/591f/2e46096d3f0330b08ce87a45/wg2020-03-03-en.pdf

Croteau, D., & Hoynes, W. (2013). *Media/society: Industries, images, and audiences*. Sage Publications.

Dados, N., & Connell, R. (2012). The global south. *Contexts, 11*(1), 12–13.

Economy, E. C. (2011). *The river runs black: The environmental challenge to China's future. A council on foreign relations book*. Cornell University Press.

Floyd, R. (2016). Extraordinary or ordinary emergency measures: What, and who, defines the 'success' of securitisation? *Cambridge Review of International Affairs, 29*(2), 677–694. https://doi.org/10.1080/09557571.2015.1077651

Friends of Nature (FON). (2018). *The friends of nature*. Retrieved from http://f-on.org/index.php/en/

Fu, D., & Distelhorst, G. (2018). Grassroots participation and repression under Hu Jintao and Xi Jinping. *The China Journal, 79*(1), 100-122. https://doi.org/10.1086/694299

Gadgil, M., Daniels, R. R., Ganeshaiah, K. N., Prasad, S. N., Murthy, M. S. R., Jha, C. S., ... & Subramanian, K. A. (2011). Mapping ecologically sensitive, significant and salient areas of Western Ghats: Proposed protocols and methodology. *Current Science*, 175–182. https://www.jstor.org/stable/24073043

Gore, C. (2000). The rise and fall of the Washington Consensus as a paradigm for developing countries. *World Development, 28*(5), 789–804. https://doi.org/10.1016/S0305-750X(99)00160-6

Guangyao, Z. (2016). *Ecological civilisation, a national strategy for innovative green concerted open and innovative development*. Retrieved from http://web.unep.org/ourplanet/march-2016/articles/ecological-civilization/

Guha, R. (2014). *Environmentalism: A global history*. Penguin UK.

Ho, P. (2001). Greening without conflict? Environmentalism, NGOs and civil society in China. *Development and Change, 32*(5), 893–921. https://doi.org/10.1111/14677660.00231

Jahiel, A. R. (1997). The contradictory impact of reform on environmental protection in China. *The China Quarterly, 149*, 81–103. https://doi.org/10.1017/S0305741000043642

Joseph, J. (2020). Exploring the agency-structure dynamics in state-society relations in contemporary China: The case of securitization of the environmental sector. *China Report*. SAGE publications. https://journals.sagepub.com/doi/10.1177/0009445520916873

Joseph, J. (2023). State, society and environmental security in international relations theory. *Fudan Journal of the Humanities and Social Sciences*, *16*(2), 171–190. https://doi.org/10.1007/s40647-022-00363-9

Joseph, J., & Karackattu, J. T. (2022). State actions and the environment: Examining the concept of ecological security in China. *Environment, Development and Sustainability*, *24*(11), 13057–13082. https://doi.org/10.1007/s10668-021-01982-0

Joseph, J., & Thomas Karackattu, J. (2023). Public protests and environmental policy-making: The cases of Xiamen anti-paraxylene protests in China and the civic movement against Kodaikanal mercury poisoning in India. *Risk, Hazards & Crisis in Public Policy*, *14*(2), 94–114. https://doi.org/10.1002/rhc3.12251

Josselin, D., & Wallace, W. (2001). Non-state actors in world politics: A framework. *Non-state actors in world politics* (pp. 1–20). Palgrave Macmillan UK.

Kahn, M. E., & Zheng, S. (2016). *Blue skies over Beijing: Economic growth and the environment in China*. Princeton University Press.

Kapur, S., & Mabon, S. (2018). The Copenhagen school goes global: Securitisation in the non-west. *Global Discourse*, *8*(1), 1–4. https://doi.org/10.1080/23269995.2018.1424686

Krahmann, E. (2005). From state to non-state actors: The emergence of security governance. *New Threats and New Actors in International Security* (pp. 3–19). Palgrave Macmillan.

Liu, D. (2013). *NGOs emerge in China, but face more challenges*. Thomson Reuters Foundation. Retrieved from http://news.trust.org//item/20131007121609-glxo2/

Liu, L., Wang, P., & Wu, T. (2017). The role of nongovernmental organisations in China's climate change governance. *Wiley Interdisciplinary Reviews: Climate Change*, *8*(6), https://doi.org/10.1002/wcc.483

Lu, P., Ren, X., Erskine, T., Guzzini, S., Buzan, B., Jahn, B., & Rosenberg, J. (2024). Debating the Chinese school(s) of IR theory. *The Chinese Journal of International Politics*, *17*(3), 277–305. https://doi.org/10.1093/cjip/poae015

Meidan, M. (2022). Energy and the economy in China. *The Palgrave handbook of international energy economics*, 631–647. Springer International Publishing.

Mendiluce, M. (2024, February 29). The green industrial race: US versus China María. *Forbes*. https://www.forbes.com/sites/mariamendiluce/2024/02/29/the-green-industrial-race-us-versus-china/

Nabhi, U. (2006). Environmental movements in India: An assessment of their impact on state and non-state actors. *India Quarterly, 62*(1), 123–145. https://doi.org/10.1177/097492840606200106

Nayak, A. K. (2015). Environmental movements in India. *Journal of Developing Societies, 31*(2), 249–280. https://doi.org/10.1177/0169796X15576172

Nyman, J., & Zeng, J. (2016). Securitisation in Chinese climate and energy politics. *Wiley Interdisciplinary Reviews: Climate Change, 7*(2), 301–313. https://doi.org/10.1002/wcc.387

Pan, J. (2016). *China's environmental governing and ecological civilization.* Springer.

People's Daily. (2004). 4th Plenary Session of 16th CPC Central Committee. Retrieved from http://en.people.cn/200409/07/eng20040907_156241.html

Sahu, G. (2008). Public interest environmental litigations in India: Contributions and complications. *The Indian Journal of Political Science,* 745–758. https://www.jstor.org/stable/41856466

Saich, T. (2000). Negotiating the state: The development of social organizations in China. *The China Quarterly, 161,* 124–141. https://doi.org/10.1017/S0305741000003969

Sanders, R. (1999). The political economy of Chinese environmental protection: Lessons of the Mao and Deng years. *Third World Quarterly, 20*(6), 1201–1214. https://doi.org/10.1080/01436599913361

Shah, G. (2004). *Social movements in India: A review of the literature.* Sage Publications.

Shapiro, J. (2001). *Mao's war against nature.* Cambridge University Press.

Sharma, M. (2024). *DALIT ECOLOGIES: Caste and environmental justice.* CAMBRIDGE University Press.

Sima, Y. (2011). Grassroots environmental activism and the Internet: Constructing a green public sphere in China. *Asian Studies Review, 35*(4), 477–497. https://doi.org/10.1080/10357823.2011.628007

Sivaramakrishnan, K. (2011). Environment, law, and democracy in India. *The Journal of Asian Studies, 70*(4), 905–928. https://doi.org/10.1017/S0021911811001719

Song, L., & Woo, W. T. (2008). China's dilemmas in the 21st Century. L. Song & WT Woo (Eds.), *China's Dilemma: Economic growth, the environment and climate change.* Brookings Institution Press.

Sriramachari, S. (2004). The Bhopal gas tragedy: An environmental disaster. *Current Science, 86*(7), 905–920. Available at http://indiaenvironmentportal.org.in/files/The%20Bhopal%20gas%20tragedy.pdf

Tickner, A. B., & Wæver, O. (Eds.) (2009). *International relations scholarship around the world.* Routledge.

United Nations Framework Convention on Climate Change (UNFCCC). (2017). *China and India lead global renewable energy transition.* REPORT/ 21 APR, 2017. Retrieved from https://unfccc.int/news/china-and-india-lea dglobal-renewable-energy-transition

Van der Kamp, D. (2021). Can police patrols prevent pollution? The limits of authoritarian environmental governance in China. *Comparative Politics, 53*(3), 403–433. https://doi.org/10.5129/001041521X15982729490361

Varghese, S. (2024). Dynamics of social media networks in the post-truth era. In S. Dahiya & K. Trehan (Eds.), *Handbook of digital journalism.* Springer. https://doi.org/10.1007/978-981-99-6675-2_36

Vuori, J. A. (2024). Chinese macrosecuritization: China's alignment in global security discourses. *Taylor & Francis* (p. 262).

Waever, O. (2008). The changing agenda of societal security. In H. Brauch et al. (Eds.), *Globalisation and environmental challenges: Reconceptualizing security in the 21st century. Hexagon Series on Human and Environmental Security and Peace book series* (HSHES, Vol. 3, pp. 581–593). Springer.

Wang, Q. (2014). Effects of urbanisation on energy consumption in China. *Energy Policy, 65*, 332–339.

Weiss, T. G., Seyle, D. C., & Coolidge, K. (2013). The rise of non-state actors in global governance: Opportunities and limitations. Available at https://oneearthfuture.org/sites/default/files/documents/publicati ons/GGWEISSFinalR6_0.pdf

Wendt, A. (1995). Constructing international politics. *International Security, 20*(1), 71–81. https://doi.org/10.2307/2539217

Xinhua News Agency, (2003). Communiqué of the Third Plenary Session of the 16th Central Committee of the Chinese Communist Party. Retrieved from http://www.people.com.cn/GB/shizheng/1024/2133923.html

Xu, J., & Byrne, J. (2021). Explaining the evolution of China's government–environmental NGO relations since the 1990s: A conceptual framework and case study. *Asian Studies Review, 45*(4), 615–634. https://doi.org/10.1080/ 10357823.2020.1828824

Xu, W., & Mao, Y. (2022). Evaluation of the industrial green transformation effect of environmental regulation: An empirical analysis based on the intermediary effect model. *Vitality, 09*, 58–60.

Yang, G. (2003a). Weaving a green web: The Internet and environmental activism in China. *China Environment Series, 6*, 89–92.

Yang, G. (2003b). The Co-evolution of the Internet and civil society in China. *Asian Survey, 43*(3), 405–422. https://doi.org/10.1525/as.2003.43.3.405

Yang, G. (2005a). Environmental NGOs and institutional dynamics in China. *The China Quarterly, 181*, 46–66. https://doi.org/10.1017/S03057410050 00032

Yang, G. (2005b). Information technology and grassroots democracy: A case study of environmental activism in China. Paper Presented at the Fourth Annual Kent State University Symposium on Democracy. Retrieved from http://upress.kent.edu/Nieman/Information_Technology.htm

Yang, G. (2011). *The power of the Internet in China: Citizen activism online. Contemporary Asia in the world*. Columbia University Press.

Young, O. R., Guttman, D., Qi, Y., Bachus, K., Belis, D., Cheng, H., ... & Zhu, X. (2015). Institutionalised governance processes: Comparing environmental problem solving in China and the United States. *Global Environmental Change, 31*, 163–173. https://doi.org/10.1016/j.gloenvcha.2015.01.010

Yu, Y., & Zeng, F. (2010). Digital power: Public participation in an environmental controversy. In J. J. Kassiola & S. Guo (Eds.), *China's environmental crisis* (pp. 179–193). Palgrave Macmillan.

Zhang, J. Y., & Barr, M. (2013). *Green politics in China: Environmental governance and state-society relations*. Pluto Press.

Zhao, S. (2022). *The dragon roars back: Transformational leaders and dynamics of Chinese foreign policy*. Stanford University Press. https://doi.org/10.1515/9781503634152

Environment and International Relations Theory

Securitisation of Development, Climate Complex and Global Climate Governance: The US, China and India

Anjan Kumar Sahu○

Abstract Despite the three decades of international climate change negotiations, there has been no substantial breakthrough on climate change pact. What is the preeminent cause and who are responsible of the present climate impasse? This chapter argues that the main factor that impedes an effective climate policy is the politics of climate change among three major greenhouse gas emitters—the US, China and India. The politics is centred on the premise of economic development as climate change policies are framed as threats to development (referent object). The framing of threat and referent object drives the major climate powers to develop a climate complex where climate policy of a country affects the behaviour of others and subsequently securitises development, particularly when the rivalry between the powers is potentially high. This study employs the Securitisation theory to examine the tripartite Securitisation of development of the major powers—the US, China and India.

A. K. Sahu (✉)
University of Allahabad, Allahabad, India
e-mail: anjanjnu@gmail.com

© The Author(s), under exclusive license to Springer Nature
Singapore Pte Ltd. 2024
J. T. Karackattu et al. (eds.), *Environmental Securitisation in India and China*, https://doi.org/10.1007/978-981-97-9160-6_2

29

Keywords Climate complex · Securitisation of development · Economic development · Securitisation · The US · China · India

Introduction

The issue of climate change is presented as a threat. The threat discourse drives international and national actors to securitise climate change. Internationally, the debate on Securitisation of climate change began on April 17, 2007, when the United Nations Security Council underscored the issue of climate change as a pressing problem which potentially threatens international peace and security. Under the Council President, Margaret Beckett, former Foreign Secretary of the United Kingdom, explicitly and emphatically, argued that climate change is a threat to international peace and security (2007). Simultaneously, some countries securitise climate change by constructing climate change as a threat (Floyd, 2010; Nyman & Zeng, 2016; Sahu, 2019).

However, despite a series of attempts on Securitisation of climate change and utmost urgency shown by some developed and developing countries (whether at the national or international levels) to address the climate issue, there has been a sheer ennui in arresting climate threats. So, this chapter presents a very pertinent and an essential research question by investigating on what is the motivating factor that drives a cluster of major and irresponsible climate powers—the United States (US), China and India—to develop a tripartite climate complex and how the complex triggers the Securitisation of development. The climate complex is the result of interlinking climate policies of the three countries who present climate policies rather than climate change as threats to their economic development. This climate complex subsequently intensifies Securitisation of development. Development is articulated as a referent object, and climate policies are constructed as threats. As the climate complex is relational, interrelated, conflicting and competitive, climate policy of one country shapes the climate behaviour of others, particularly when the competition and rivalry between/among major powers are high as we see the evolving rivalry between the US and China. The deeply ingrained potential threat to economic progress prevents them to reach a concrete global climate change treaty, though the triad undertakes a slew of domestic climate measures where the prospect of economic gain is high. This

Securitisation of development is called miniSecuritisation as development is securitised by a cluster of powerful and influential countries—Washington, Beijing and New Delhi. It is worthwhile to mention that the objective of the chapter is not to examine the Securitisation of development and the role of significant actors in setting Securitisation in their respective domestic climate policy. There has been a good number of research on the process of Securitisation and the role of securitizing actors in determining domestic climate policies of the US, China and India (Floyd, 2010; Nyman & Zeng, 2016; Sahu, 2021, 2022). However, it is crucial to mention that domestic factors are indispensable to set the Securitisation of development beyond the territorial boundary.

The debate on Securitisation of development has been a recent development in the Western countries' aid policy to developing countries (Bagoyoko & Gibert, 2009; Fisher & Anderson, 2015). One of the adverse consequences of the aid policy is that the Western countries, particularly the US, candidly support and strengthen militaries of 'illiberal states' of developing countries—mostly African countries—that securitise the development processes and enhance the unrestrained role of military and political elites (Fisher & Anderson, 2015). However, the Securitisation of development in the environment literature is largely unexplored, except the recent debate on Securitisation of development and climate change interrelationship in the context of India and China (Sahu, 2021, 2022). However, the study of China and India primarily explains the domestic Securitisation of development. This chapter takes the debate forward to shed light on the inter-states' Securitisation of development— the US, China and India. It means China and India's climate policies are perceived by the US as a threat to its economic interests and vice versa.

These tripartite states are critical for five crucial reasons. First, the triad are the first (China), second (the US) and the third largest (India) greenhouse gas emitters in terms of cumulative emission. Unless they collectively agree on a common climate treaty, any attempt of containing the climate danger would be ineffective or futile. Second, both the US and China compete to get maximum advantage in the global climate arrangement and they know the importance of India as New Delhi is the most populous country which has the potential to define the course of the global climate governance. This was reflected both in Copenhagen climate treaty, held in 2009, that compelled the Chinese Premier Wen Jiabao to seek the support of Indian government to counter the US proposal on a legally binding emission agreement and applying the potential mechanism of 'international analysis and assessment' to verify the national climate change measures of countries, including developing states (Saran, 2019). The US also cannot ignore New Delhi's prominent role in formulating a global climate pact as it is evident in the Paris conference (Lavasa, 2019). For example, PM Narendra Modi's enunciation of International Solar Alliance in the Paris conference demonstrates New Delhi's increased international stature to stem climate crisis. Third, the evolving global geopolitical rivalry, particularly China's potential threat to the US-dominated 'reginal security order' in East Asia and the Indo-Pacific, might determine global climate governance and India is the major actor to influence the global rivalry (Buzan, 2020; Saran, 2019). The integration of rivalry politics with climate change is inseparable. Fourth, the triad has sufficient economic and political clouts to shape the global climate discourse. It is evident as the US denies accepting a legal climate pact unless China and India accept legal obligation and vice versa. Lastly, the rising economic profiles of China and India restructure the North–South dyad as the two countries' economy substantially determines international political economy. The rising economic profile of Beijing and New Delhi is a cause of worry for the US as economically powerful countries determine global political and economic governance, and slacken the long-standing influence of the US. Against this backdrop, the formation of a climate complex and Securitisation of development is a novel way of analysing the climate politics that has implications for global climate governance. The whole idea of climate complex and Securitisation of development is explained by employing the Securitisation theory that underscores the role of state actors in defining external climate policy that triggers climate dilemma, conflict and judgement of actors involved in the process. According to Buzan et al. 'it *does* matter how others judge the reasonableness of a Securitisation, because this influences how other actors in the system respond to a security claim' (1998: 30). Thus, how these three climate actors respond and perceive the climate policy is significant to determine the Securitisation of development.

This paper is structured as follows. Apart from the introduction, the second section sets out a theoretical debate on Securitisation theory and its linkage with the Securitisation of development. The third section explicates how the prominent climate players securitise their development at the interstate level that leads to a climate complex. The last section highlights the main findings of this paper.

Theory of Securitisation and Securitisation of Development

The theory of Securitisation is a novel way of understanding the process of framing a particular issue as a threat and a referent object. It means the identification of a threat and a referent object is deliberately constructed. Protection of a referent object from a threat is imperative as an unprotected and unsecured referent object unleashes massive social, economic and political upheavals which may cause the collapse of a system. Thus, a set of necessary and urgent measures must be undertaken to prevent the threat before it becomes uncontrollable. The threat-defence logic is decided by securitizing actors—primarily influential political leaders and bureaucrats—and the logic is accepted by the 'relevant audience' or general public (Buzan et al., 1998).[1] Once the audience extend their support to the securitizing actors' designation of threats and referent objects, Securitisation process is successful (Buzan et al., 1998). However, Securitisation of an issue has both internal and external implications. At the national level, Securitisation triggers undemocratic repercussions such as violation of people's rights and liberty. At the interstate or global level, Securitisation of an issue dissuades countries to cooperate and drives them to embroil in conflict centred with 'fear, perception and uncertainty' or security complex (Sahu & Mohan, 2022). This paper deals with the Securitisation of development seeing the external threat posed by a rival or contender to outwit others, though domestic factors are not completely discounted because Securitisation of development at the domestic level affects the foreign policy of a country. However, due to the space constraint, this chapter examines the external climate policies of the major climate powers and how their climate policies are interconnected that deepens Securitisation of development loaded with uncertainty, fear, misperception and security complex.

However, over the years, it is noticed that scholars have extensively examined the theoretical and empirical issues of Securitisation covering traditional and non-traditional security challenges such as health, climate and development (Elbe, 2006; Floyd, 2010; Sahu, 2021). This chapter

---

[1] Relevant audience depends on the context. Sometime, relevant audience are people in the inner circle of the government or actors that support the stance of political actors that excludes the general public (see Sahu, 2022). Sometimes, Securitisation depends on the acceptance of the threat-defence argument by the general public (Hayes, 2012).

unpacks the linkage between security and development to analyse climate change discourse. However, it is worth noting that the security and development interrelationship is not independent from other sectors like national security, environment, political rivalry, etc. The relationship between security and development is understood by looking into other sectors. However, this chapter argues that the security-development narrative has an edge over other narratives in examining the Securitisation of development. Against this setting, it is essential to briefly underline the security-development nexus and how Securitisation steers inter-states' climate policy.

The nexus between security and development is not a novel one. In the post-Second World War, the linkage occurred to determine global politics, particularly in the context of the Cold War. In consonance with military security, the idea of the modern concept of development used as a major foreign policy to influence the global political economy. To cement its geopolitical, economic and ideological interests, Harry S Truman underlined the problem of economic underdevelopment in underdeveloped countries. Terming communism as a 'false philosophy', Truman invoked the urgency of economic development, democratic values and 'democratic fair-dealing' to redress underdevelopment in underdeveloped countries (Truman, 1949). Though political or national security was not explicit in his address, Truman's agenda was to deploy the normative order based on development needs of the people to serve its security and economic interests in order to counterbalance and curtail the prominence of the Soviet Union in the economically underdeveloped countries. However, in the post-Cold War period as the idea of security was widened and deepened, development became an integral part of a security discourse. Apart from the negative side of development-security nexus, it is also observed in the post-Cold War phase policy-makers and donor agencies provided aids and were able to contain violence by isolating violent actors and stabilising relation between competing groups (Anderson, 1996, cited in Duffield, 2010). However, the emergence of concepts like human development and human security further buttressed the security-development nexus in contrast to the state and economic-dominated development. The 'liberal way of development' concerned with meeting basic needs, education and 'adaptive self-reliance' indicates the growing significance of human aspect of development (Duffield, 2010: 55).[2]

However, the intermingling of security-development drove some scholars to analyse the Securitisation of development that enhances overwhelming role of state-centric security discourse. The Securitisation of development idea was dominant in the post-9/11 global security order as some Western countries developed the discourse of 'global war on terror' and adopted the method of supporting certain countries in the form of economic aid, particularly in the 'hot spot' areas affected by the scourge of terrorism (Fisher & Anderson, 2015). However, it is observed that the Securitisation of development does not solely originate from the Western countries (Fisher & Anderson, 2015). Some southern countries deliberately embraced Securitisation in their respective region or states. By getting unswerving support from the Western countries, some of the governments in Africa securitised development by overwhelmingly exercising state power and militarizing policy (Fisher & Anderson, 2015).

However, the connection between Securitisation and development is a novel perspective in the field of environment literature. Recently, climate change has been a 'security priority' (Dalby, 2015: 427) or articulated as a major threat to international peace and security (UNSC, 2007). However, what complicates the present climate and security discourse is the interrelationship of climate, security and economic development. In the context of Securitisation of development, development is the referent object and the nature of a threat connected with the object. The nature of a threat may be military or non-military. In the context of a military threat, non-military aspect is not inseparable. Likewise, a military issue is not unconnected from a non-military threat. It means the relationship between the military and non-military threats cannot be rigidly categorized. For example, if a nation is militarily threatened, then the economic aspect of the military threat cannot be ignored. Similarly, if a country faces an economic crisis, then its implications for military security is unavoidable. Here, the paper analyses the framing of a climate policy as a threat to the economic development and an unharmed economy assists to enhance military security.

---

[2] 'In distinction to modernization and economic catch-up strategies, the key characteristic of the liberal way of development is the privileging of notions of sustainability based upon adaptive patterns of household and communal self-reliance in the global south' (see Duffield, 2010: 55–56).

The debate on Securitisation of development is a new phenomenon in the climate security discourse (Sahu, 2022). It means economic development is designed as a referent object and climate policy is posed as a threat to the object. Political leaders and policy-makers underline the imperative of development that drives the existing global climate politics in their favour, and employ military means or economic logic to legitimise their claims. Thus, the economics of climate change has an edge that significantly shapes the politics of climate change around global climate negotiations. It is worth mentioning that the economic development idea is projected as a normative imperative. The normative nature of economic development is connected with climate policies and people's development. The recurring claim on the development of the people and enhancing their choices in the climate discourse emboldens political leaders and policy-makers to legitimise their move to securitise development, though the rationale behind the development logic is against the ethos of environment and interest of the people (Sahu, 2022). The normative imperative has both internal and external dimensions to legitimise the Securitisation of development. Internally, countries refrain from accepting climate responsibility by invoking the development gap or unchanged lifestyle of its people. Externally, putting people's development at the centre, countries do not want to address the problem. It means countries employ domestic and international factors to strengthen their normative agendas. Thus, the idea of Securitisation of development is connected with climate change, military objectives and normative agendas. However, the normative ideas are not inseparable from a non-normative climate policy as Securitisation of development emphasizes the urgency of economic development rather than redressing climate dangers. The following sections underline how these three major emitters securitise their development in their climate discourse.

SECURITISATION OF DEVELOPMENT: THE US, CHINA AND INDIA

The idea of Securitisation has both domestic and international dimensions. This paper does not deal with the domestic aspect as there has been already an established literature on how climate change or development is securitised in the US, China and India, and the prominent actors' involvement in the process of Securitisation (see Floyd, 2010; Nyman & Zeng, 2016; Sahu, 2021, 2022). However, while examining the climate discourse of the triad, domestic discourse debate is employed to support and substantiate their Securitisation of development at the inter-states level.

In the previous section, it is underlined that securitizing actors designate economic development as a referent object and normative discourse is employed in legitimising the idea of development. The origin of the process can be traced to the commencement of the Stockholm Conference in 1972. In the conference, developing countries'—particularly, China and India—demand for economic development rather than environmental protection substantially accepted as a norm in the subsequent international climate change negotiations. Without fully relegating environmental issues, economic development was given utmost priority seeing the abject poverty of the developing countries. The United Nations Framework Conventions on Climate Change (UNFCCC) and the Kyoto Protocol are the twin cases that legally put no restriction on developing countries for committing emission reduction. From the perspective of developing countries, a non-obligatory climate treaty is essential to ameliorate the large number of poor people's standard of living as obligation to cut emission would seriously undercut the people's survival and security. At the Beijing Ministerial Conference on Environment and Development of Developing Countries in 1991, China strongly defended developing countries' right to development and access to markets of developed countries, and advocated for responsibilities of the developed countries for addressing environmental degradation and providing financial and technological support to developing countries (Jiang, 2022). Likewise, India's policy-makers and prominent NGOs associated with the Government of India—such as Center for Science and Environment, and the Energy and Resources Institute—firmly put the developmental rationale of the country (Sahu, 2022). It means China and India deployed the development imperative and normative rationale for their exemption from the legal pact. China and India's consistent push for the development needs of developing countries drove the UNFCCC to codify the binding and non-binding climate change pact that supported the developing countries' development imperative and instructed the developed countries to adhere to a binding emission reduction commitment. The 1992 UNFCCC agreement obliged the developed countries to accept a binding commitment to reduce emission and exempted developing countries from the obligation as the former is historically responsible for the global climate change problem and framing the developing countries as non-obligatory seeing the development progress of the North and the South (UNFCCC, 1992).

However, UNFCCC and Kyoto Protocol received a major setback as the US—the largest cumulative and per capita emitter in the 1990s—refused to be a part of the two treaties and the Congress, legislative branch of the country, declined to ratify these twin climate pacts. What prompted the US to take this stand is the unwillingness of political leaders to change the lifestyle of its people as binding climate emission would compel the people to reduce their carbon emission and consumption pattern. Another factor is refusal of some economically growing countries especially China and India to accept the emission reduction commitment. The Byrd-Hagel Resolution is a case in point. Though the Resolution has domestic angel, external factors are unavoidable in support of it. 'Yet the most significant change in the political terrain came with the Byrd-Hagel Resolution, which the US Senate passed 95–0 on July 25, 1997, four months before the Kyoto negotiations commenced'.

However, the official stand of the US on global climate negotiation goes back to the 1990s when the then President of the US H. W. Bush signed the UNFCCC in 1992 stating that 'the American way of life is not negotiable' (cited in Harris, 2000: xxv). Focussing on the economic dimension of climate change, Robert Reinstein, head of the U.S. delegation to the Framework Convention negotiations, stated that 'The issues are at the heart of the economy-they are extremely complicated' (cited in Betsill, 2000: 214). Apart from domestic audience and China-India factor, influential businesses such as fossil fuel and automobile industries strongly influenced international climate position of the US through the Global Climate Coalition (Bestill, 2000).[3] White House Chief of Staff John Sununu, an important official managing the issue of climate change during J. W Bush regime, maintained that decarbonisation is more about 'no growth policy' rather than containing 'a valid environmental threat' (Bestill, 2000: 215). The non-binding resolution resolved that the US should "not be a signatory to any protocol' unless it mandated specific commitments for developing countries, *and* would 'not result in serious harm' to the US economy' (Downie, 2013: 30). Subsequently, the former President of the US G. W. Bush withdrew support from the Kyoto

---

[3] The Global Climate Coalition (GCC) was formed in 1989. Its members include the American Petroleum Institute, Chevron, Chrysler, General Motors and the National Mining Association.

Protocol which was later not ratified by the country and 'placed economic interests at the heart of its new foreign policy' (Falkner, 2005: 592). Just few months after becoming the President of the US, G. W. Bush opposed the Kyoto Protocol by saying that the Protocol 'exempts 80 percent of the world, including major population centers such as China and India, from compliance' and the potential 'serious harm to the US economy' (cited in Harrison, 2007: 104). Further he said: 'I will not accept a plan that will harm our economy and hurt American workers. Because first things first are the people who live in America. That's my priority' (cited in Harrison, 2007: 104). The non-inclusion of emerging economies in the binding climate fact even induced Donald Trump's decision of withdrawal from the Paris Agreement (Eckersley, 2022). According to Falkner, 'Ever since the United States took a backseat at the UN Conference on Environment and Development (UNCED) in 1992, US foreign policy has appeared to be lukewarm about, and often hostile to, multilateral environmental policy-making. From the rejection of the Convention on Biological Diversity (CBD) to the withdrawal from the Kyoto Protocol on climate change, the United States has shown itself to be concerned more with national economic interests than global environmental threats' (2005: 585). Seeing the current climate position of the US, Eckersley maintains that the US is reluctant to address climate issues as multilateral climate change agreements threaten the country's 'relative economic advantage under the existing international order' (2022: 52).

However, in the first decade of the twenty-first century, a rapid economic progress of China and India, and the US concern of the potential threat from them to weaken its economy led to a sharp global climate dyad: the Global North and the Global South. Since the introduction of the UNFCCC provisions in 1992, China and India relentlessly stick to their developing countries status to studiously avoid any legally binding commitment. However, in the last couple of decades, the two Asian powers effectively compete with the developed countries—economically, militarily and politically—by projecting their status and strength. The major factor that drives the two Asian powers to augment their great power status is the rising economic profiles.

Despite the growing economic clout, Beijing and New Delhi refute to be parts of legal climate commitment arguing that the people in their respective countries still face development gap—energy inaccessibility, inability to meet basic needs, inability to deal with climate adaptation, dearth of financial and technological support. Industrial development is the panacea for meeting the development needs of the people. Cutting across party lines, political leaders and officials persistently stress on the need for economic development, though the domestic compulsion for economic development varies between the two countries. India claims that 'It not only has to complete the current unfinished development agenda, it has to strategise for future pressures that may increase the magnitude of this development gap' (INDC, 2015).

However, both India and China advocate the imperative for economic development as the objective is aligned with the other third world countries. Presenting the demand for development, Beijing and New Delhi argue that the present development crisis is the outcome of historical emission of developed countries, and the latter has the moral and legal responsibilities to redress the development deficit in developing countries by providing adequate climate fund, accepting its historical mistake, transferring sophisticated climate mitigation technology, recognising climate equity and advocating per capita emission. It means Beijing and New Delhi shift the blame to developed countries, including the US, without accepting a legal climate pact. Koehn observes that: the position on climatic-change policy articulated by PRC officials involves a mix of perceived national self-interest and commitment to the worldview shared by many countries in the Global South...Mainland leaders are able to justify continued pursuit of their chosen growth-dependent path without incurring pressures for immediate emission-mitigation actions from poor countries that value China's leadership in advancing the South's agenda (2008: 55–56).

As the two potent economic forces are intransient in their climate stance, the US perceives this intransience is a deliberate strategy of China and India to outwit Washington economically which has implications for enhancing the status, political power and military superiority as economic development substantially assists strengthening the power of a state in all its dimensions. The US political leaders had developed this understanding even before the conclusion of Kyoto Protocol in 1997. In the first decade of the twenty-first century, the US is more concerned about its possible acceptance of climate change pact that might threaten its economic development and provide an opportunity to both Beijing and New Delhi to economically outstrip its economy. The US's dilemma of Beijing and New Delhi's potential of surpassing the economy of the former compels Washington's political and business elites to refrain from a compulsory emission reduction treaty. The global economic crisis in 2008 further induced the policy-makers of the US to undertake measures to stabilise national economy rather than showing concerns for global climate change.

Unlike China and India, one of the factors for stressing the vitality of development is a cluster of powerful groups or 'climate *deniers*'—citizens, businesses, officials and legislators—shape the climate policy of the country by discrediting climate science, and espousing the personal and economic freedom as obligation to act on climate change undermines or restrains the long-held belief in a free-market ideology (Harris, 2000). This argument is centred on the principle that legal obligation prevents unhindered economic and market policies of the government which has been the central pillar of the country's economic policy since the Second World War, especially during the Cold War phase that divided the world on market and socialist ideologies. Despite the preferences for an economic ideology and business as usual practice, Washington consistently presses for its long-standing position in the international climate negotiation—to 'break the calcified distinction between developed and developing countries that had always characterized climate change negotiations' (Harris, 2000: xxiii). However, what is common among the three is the continuing dependence on fossil fuels to power their industries. The economic perspective of climate change deeply became an international norm as many countries internalised that climate change is a threat to their economic development, particularly the period from 1988 to 1992 that shifted the climate change problem from an environmental crisis to an economic danger as the acceptance of climate obligation would impose huge economic burden or cost on industrial as well as non-industrial countries (Betsill, 2000). The preoccupation with the economic aspect is strongly embedded in the policies of Beijing, New Delhi and Washington with the support of all major political parties.

It is true that Beijing and New Delhi are prominent players in global climate change governance; however, their policies are primarily guided by their economic motives, stemming from the climate business. For example, the introduction of Clean Development Mechanism (CDM) provided a massive opportunity to China to gain technology as 56 per cent of the CDM projects located in Beijing. The CDM provision widely helped Beijing in transformation of solar and other renewable energy that broadly serves the strategic and economic interests. MacNeil and Paterson observe that 'the CDM contributed to, alongside the broader transformations in China's political economy and the powerful domestic health and air pollution reasons to address China's coal use, a shift in orientation where China started to see its strategy as consistent with a shift to becoming a leader in both the manufacturing and installation of renewable energy. The result was that rapidly, China became the world's largest solar pV manufacturer and exporter, with a 60.4% share of global production in 2013' (2019: 6–7). Another instance is when the German government proposed the idea of forming a Renewables Club in 2013, the governments of China, India, Morocco, South Africa, Tonga and the United Arab Emirates as non-EU members declared their interest in joining (Hovi et al., 2016, cited in Tosun, 2018). In the case of China, India and South Africa, the decision to join the Renewables Club can be ascribed to their recognition of the EU's market power and the potential risk to their economic development stemming from carbon-dioxide-intense industrialisation (Tosun, 2018). New Delhi's initiative for International Solar Alliance is laudable by some scholars as the country has shown its climate leadership. However, the country remains outside of the mandatory emission reduction commitment and the Alliance is largely guided by the country's economic motive as the rising renewable energy markets may determine the global markets in the coming decades. Even China and the US are not unaware about the booming renewable energy markets.

However, the consistent international pressure on China and India forced these countries to accept a voluntary climate change measures to reduce emission. The triad were elated by the 2015 Paris agreement (Twenty-First Conference of the Parties [CoP-21]) which for the first time instructed major countries, including the triad, to stick to climate policy obligation (Dimitrov, 2016). However, the US is aware about the fact that voluntary climate measures argument may not guarantee an assured reduction of emission and Washington may face political and public criticisms if the country accepts a binding climate pact. In the Paris Conference, US political elites opposed the legally binding agreement and China, though favoured an obligatory treaty, refused to accept the proposed review of developing countries' climate policy measures. A top European Union official says, 'If we insist on legally binding, the deal will not be global because we will lose the US' (Dimitrov, 2016: 3). Likewise, an official attending an internal climate change meeting in the Paris Conference mentions that 'China is maximalist on legally binding and minimalist on transparency' (Dimitrov, 2016: 3). Thus, the economic development factor grossly guides the climate behaviour of the three prominent climate players.

The triad perceives each other's climate policy as a threat to their economy. The US policy of insisting the inclusion of China and India in the legally acceptable climate change actions, refusal to accept historical emission, disregard for climate inequity and ignoring the development needs of the latter are perceived as threats by Beijing and New Delhi. In contrast, Beijing and New Delhi's refusal to accept emission reduction commitment despite their rising emissions, growing economic profile, high level of intra-national climate inequity and potential to surpass the economy of Washington (which has implications for the military) is perceived by the US as threats to its economy. It means both China-India and the US climate behaviour is based on mutual perception and judgement. This interrelated climate perception germinates distrust, competition, conflict and mutual suspicion. This is called climate complex as the development of this complex drives them away from a global climate cooperation. The trilateral climate complex is guided and determined by the ingrained notion of Securitisation of development as development is given the supreme priority in their climate policy.

CONCLUSION

Constructing climate change within the economic domain is not a recent phenomenon. The economic aspect of climate change emerged in the 1970s as underdeveloped and developing countries underlined the imperative of economic progress to redress impoverishment. However, what is new is the intensification and interrelated nature of economic narrative that substantially shaped the climate policies of some major and powerful climate emitters. The interrelated nature of economic dimension led countries—the US, China and India—to formulate their climate policies seeing each other's climate actions. It means climate policies of the US are connected with the policies of China and India and vice versa. The interconnected and relational character of climate policies forced them to develop a mutual and tripartite climate complex centred on the Securitisation of development. The climate complex is not immune from geopolitical rivalry, especially between the US and China. As the rivalry is getting endured and Beijing is a potential security challenger to Washington, the latter is not eager to accept a legally binding climate pact that might undercut its economic choice and prospect of economic development. This climate complex is competitive, conflictual, non-cooperative and clash of economic interests.

One of the important implications of the triad's consistent attempt to securitise development is the evolving nature of the referent object that is getting incremental recognition of the wider audience and political leaders at the global level as development is presented as a symbol of prosperity of the people and the country. Consistent with their development-driven climate policy at the domestic level where development is framed as a referent object and within the territory of the state, the triad has created the discourse of development at the interstate or micro-level and deterritorialised the Securitisation of development. Being the first (China), second (the US) and third (India) largest cumulative greenhouse gas emitters, this trilateral Securitisation of development can be described as miniSecuritisation. However, if these climate powers (negatively) continue to stick to the same policy, then the miniSecuritisation would be broadened by securitizing development within the macro-Securitisation framework. It means the idea of Securitisation of development has the potential to engulf other rising and competing powers at the global level. This would be more dangerous and serious detrimental to a global consensus on addressing climate change as more countries would prefer to securitise their development taking cues from the largest emitters. To address this problem, the US, China and India must come together and set an immediate time line for a legally binding climate pact. As the largest emitters, their participation is indispensable to halt climate danger.

REFERENCES

Bagoyoko, N., & Gibert, M. V. (2009). The linkage between security, governance and development: The European Union in Africa. *The Journal of Development Studies, 45*(5), 789–814.

Beckett, M. (2007). The case for climate security. *The Royal United Services Institute (RUSI), 152*(3), 54–58.

Betsill, M. M. (2000; 2016). The United States and the Evolution of International Climate Change Norms. In P. G. Harris (Eds.), *Climate change and American Foreign Policy*. Palgrave Macmillan.

Buzan, B. (2020). Afterword to: 'Networking hegemony—alliance dynamics in East Asia'. *International Politics, 57*(2).

Buzan, B., Waever, O., & de Wilde, J. (1998). *Security. A new framework for analysis*. Boulder-London: Lynne Rienner.

Dalby, S. (2015). Climate geopolitics: Securing the global economy. *International Politics, 52*(4).

Dimitrov, R. S. (2016). The Paris agreement on climate change: Behind closed doors. *Global Environmental Politics, 16*(3).

Downie, C. (2013). Three ways to understand state actors in international negotiations: Climate change in the Clinton years (1993–2000). *Global Environmental Politics, 13*(4).

Duffield, M. (2010). The liberal way of development and the development and security impasse: Exploring the global life-chance divide. *Security Dialogue, 41*(1), 53–76.

Eckersley, R. (2022). Great expectations: The United States and the Global Environment. In R. Falkner & B. Buzan (Eds.), *Great powers, climate change, and global environmental responsibilities*. Oxford University Press.

Elbe, S. (2006). Should HIV/AIDS be securitized? The ethical dilemmas of linking HIV/AIDS and security. *International Studies Quarterly, 50*.

Falkner, R. (2005). American Hegemony and the global environment. *International Studies Review, 7*.

Fisher, J., & Anderson, D. M. (2015). Authoritarianism and the securitization of development in Africa. *International Affairs, 91*(1), 131–151.

Floyd, R. (2010). *Security and the environment*. Cambridge University Press.

Harris, P. G. (2000; 2016). Climate change and American foreign policy in the twenty-first century. In P. G. Harris (Ed.), *Climate change and American foreign policy*. Palgrave Macmillan.

Harrison, K. (2007). The road not taken: Climate change policy in Canada and the United States. *Global Environmental Politics, 7*(4).

Hayes, J. (2012). Securitization, social identity, and democratic security: Nixon, India, and the ties that bind. *International Organization, 66*(1), 63–93.

India's Intended Nationally Determined Contribution (INDC). (2015). Working towards climate justice, 6 2015. https://www4.unfccc.int/sites/ndcstaging/PublishedDocuments/India%20First/INDIA%20INDC%20TO%20UNFCCC.pdf

Jiang, C. (2022). Revisiting "leadership" in global climate governance: China's normative engagement with the CBDRs principle. *The Chinese Journal of International Politics, 15*(2).

Koehn, P. H. (2008). Underneath Kyoto: Emerging subnational government initiatives and incipient issue-bundling opportunities in China and the United States. *Global Environmental Politics, 8*(1).

Lavasa, A. (2019). Reaching agreement in Paris: A negotiator's perspective. In N. K. Dubash (Ed.), *India in a warming world integrating climate change and development*. Oxford University Press.

MacNeil, R., & Paterson, M. (2020). Trump, US climate politics, and the evolving pattern of global climate governance. *Global Change, Peace & Security, 32*(1), 1–18.

Nyman, J., & Zeng, J. (2016). Securitization in Chinese climate and energy politics. *Wires Climate Change, 7*(2), 301–313.

Sahu, A. K. (2019). The democratic securitization of climate change in India. *Asian Politics & Policy, 11*(3), 438–460.

Sahu, A. K. (2021). From the climate change threat to the securitisation of development: An analysis of China. *China Report, 57*(2).

Sahu, A. K. (2022). Referent object, securitising actors and the audience: The climate change threat and the securitisation of development in India. *Cambridge Review of International Affairs, 35*(1).

Sahu, A. K., & Mohan, S. (2022). From securitization to security complex: Climate change, water security and the India–China relations. *International Politics, 59*.

Saran, S. (2019). One long day in Copenhagen. In N. K. Dubash (Ed.), *India in a warming world integrating climate change and development*. Oxford University Press.

Tosun, J. (2018). Diffusion an outcome of and an opportunity for polycentric activity? In A. Jordan, D. Huitema, H. V. Asselt, & J. Forster (Eds.), Governing climate change: Polycentricity in action? Cambridge University Press.

Truman, H. S. (1949). Inaugural address. https://www.trumanlibrary.gov/library/public-papers/19/inaugural-address

United Nations Framework Convention on Climate Change (UNFCCC). (1992). https://unfccc.int/resource/docs/convkp/conveng.pdf

United Nations Security Council (UNSC). (2007). *Security council holds first-ever debate on impact of climate change on peace, security, hearing over 50 speakers.* https://press.un.org/en/2007/sc9000.doc.htm

CHAPTER 3

Problematic of Ecological Routines: Securitisation and Beyond

Mathew A. Varghese

Abstract There are proliferating contexts in India where one can dwell on the discourse of 'securitisation' in general and the drawing in of ecology in particular. These may be obvious steps away from knee-jerk state security narratives. This is thanks to prioritisation of ecological or environmental parameters. In addition, there is a clear step away from state as reference into others like ecosystems. But the concern is how this gets done and in what concrete state contexts. This paper tries to extend the possibility of speech as a political act in unravelling the contemporary incorporation of the environment/ecologies in state-talks/ decisions. It focusses on contexts in India as well as elsewhere that draws on exception (e.g. SEZs) and abstraction (Compensation/ Economisation). It asks when and how the state routines on ecologies become problematic. Further, it ponders whether ethnographies from ecologies in the making offer other ways of seeing/ hearing/ knowing/ theorising.

M. A. Varghese (✉)
School of International Relations and Politics, Mahatma Gandhi University, Kottayam, Kerala, India
e-mail: matzwordz@interdisciplinarity.org; m.varghese@mgu.ac.in

© The Author(s), under exclusive license to Springer Nature   47
Singapore Pte Ltd. 2024
J. T. Karackattu et al. (eds.), *Environmental Securitisation in India and China*, https://doi.org/10.1007/978-981-97-9160-6_3

Keywords Anthropocene · Compensatory · Exception · Non-Human · Risk · Securitisation · State

INTRODUCTION

There is a proliferation of mixtures of 'natural disasters', climatic events and social contradictions across India. Landslides in Himalayas, earthquake scenarios in Gujarat and elsewhere, air quality index dips in metropoles, floods in Assam, Kerala or the urban regions like Chennai, cyclones and sea-rises: everything has invoked events and contradictions of distinct kind. But underlying the responses could be seen as an overlapping narrative and call, whereby a disaster relief institutional ecosystem has to gear up. The disaster management authorities that came into place by 2005 are precisely the start of institutional response mechanisms that lay out plans, policies and designs. There is a clear line of hierarchical action that operates from the union governmental structures, through the states to district-level authorities and personnel in place.

The Himalayas off late has surely been animated by disasters that turn visible during naturalised pilgrimages or civil works on roads that chequer these young mountain ranges. The *char-dham*-like pilgrimages that have invented traditions, the option-less urgent urbanisation, as well as the damming for irrigations have all been predominantly presented as sites of unfortunate causalities rather than as ontological pointers. The Rockefeller foundation recently chose Surat, Chennai, Pune and Jaipur as part of initiatives that promote resilience in the face of disasters. Climate changes and best practices in its face are marked out and supported. Floods in Chennai and quakes in Surat were highlighted. A gamut of floods in Chennai between 2005 (when the National Disaster Act got passed) and 2015 added rationale to gear up and become resilient. Actionable initiatives, roadmaps and stakeholderships were suggested ever more in urban plans and policymaking.

Ethnographically familiar for this researcher were the flood events and the ecologies of fear and risk that ensued, in Kerala. The massive floods of 2018 that affected mostly life in central Kerala and those in the following year that mostly affected the northern parts of Kerala are cases in point. The former affected more than 5 million humans, many million non-humans as well as shattered the naturalised coordinates of everyday life.

It was an event in the sense that one could clearly mark an ante- and post-diluvian psycho-sociality. Flood risk added on to the existing risk factors like disease, economy and employment. With expectations from the state historically high, populations expected ever more from the state in terms of welfare.

The deepening of broadening of the idea of security is built well around such climatic and ecological events. Of course, with global resonances, the idea of security has slowly, but surely migrated well beyond the sole domain of state and that of war/threat of this. This is where Copenhagen school as well as post-cold-war nuances meet the political terrains in the making across India (as well as many other geopolitical locations). Securitisation theory in general and the introduction of a broad category of greening in particular in the sense of an ecologisation or environmentalisation of politics find its clear moorings. This is more so in a time when state as one knew it may be fast undergoing a morphological shift.

It is not that the traditional security challenges, like an aggressor state or terror groups, have ceased to be important. In fact, they have broadened to include any actors who may pose a challenge to the neoliberal order. The ambit of immunological responses of the state has broadened. It is into this space that ecological risks and climatic challenges that logically invoke greening/ environmentalising solutions have come in. One may point at a flood, a landslide or a sea-rise as objective threats. Not only that these objective threats constitute as security threats because it is a threat to survival. Things appear obvious at this point!

But then we have to continue by asking questions like who does this labelling and at what juncture. Securitisation theory says that only by getting labelled will an issue become a threat. So, interpellations matter and securitisation also has a potential to be extended to a discourse analysis by asking who calls what a threat as well as in understanding how the audience accepts and gets ready to elevate this threat to an exceptional status. The critical schools that take a more neo-Gramscian approach (Booth, 1991; Wyn Jones, 1996) take a further step and see the insecurities emanating from the state as well. Surely candidates on that count have proliferated in India in terms of accumulations by dispossessions, withdrawal of remaining welfare, thought policing, communalisation or shock policies like demonetisation. Here the potential of critical school as an analytical category remains to be explored.

Take the case for river-linking made by the states from the nineteenth-century colonial state with the then irrigation engineering logic to the much more vocal and comprehensive arguments made by the twenty-first-century neoliberal state in 2024. On the one hand, we see the grand entry of environment into the discourses of state. The case of human rights and water scarcities are also put together. If one closely reads the different components in the plan, from the National Water Development Authority of National Institute of Hydrology's hydrological data (knowledge), the language of infrastructural work needed, the authoring of needs and solutions to the final policy, traditional security arguments that are state centred are well in place and in more elaborate and entrenched ways that it was during colonial times. In addition, there is the lack of knowledge transparency on river basins, for example when these are spread across state boundaries. They automatically become state secrets. What is presented as excess and dearth in the context of river interlinking plans are even otherwise top-down visions with no backing of long-term ecological science. On top of this is the secretive nature of objective data presented before people with simulated ecological information upon which water securitisation arguments are built upon.

Even on a cursory gaze the ongoing green plans as well as environmental policies do not resonate with the basic premises of moving away from state-centric routines. It is even further away from the emancipatory logic of critical schools that have a mutualist and holistic idea of emancipation. In fact, they are particularly trenchant in the critique of state as provider of any form of security and their contrary position of state as one of the predominant causes of insecurity. An analytical point that one can take away from securitisation approaches for evaluation against the greening plans in perspective is the significant focus on the relationship between the classifier and the classified. The politics of security has to be part of the practice of securitisation (Floyd, 2007). In the gap of the potential for theories of security, whether those are of the Copenhagen or Critical Theory varieties, I place existing policy prescriptions that environmentalist politics structure scenarios that necessitate a rethinking of traditional notions of security and thenceforth offer solutions. The chapter explores the scope of such non-traditional understandings of security to analyse greening as incorporated in policies. It evaluates the possibilities and ponders whether there are further gaps to be addressed.

I stress three of the scenarios through which ecological solutions emerge as routines in India. The first one is along the lines of risk and risk responses exemplified by National Disaster Relief (NDR) measures after the act passed in 2005 as well as the sense of urgency built into smart city solutions to discourses of urban crisis ever since 2015, the year in which the Smart City Mission (SCM) got launched. We clearly see the hierarchical procedures involved. These proceeds through central, state and district levels without taking into account local responses in the case of the former. With respect to smart cities nuanced relationships on the ground or even the locally elected bodies become less relevant as against apolitical belief in techno-solutionism and private investments. There is a gamut of exceptionalities embedded in smart solutions that abstracts any place as a site of investment.

Post 2000, rapid scale urban processes, decentralisations, public-private partnerships, questions over designation of Special Economic Zones or creation of new urban neighbourhoods, conveys a sense of urgency. Such has been the case from the urban projects like the agglomeration plan under the Jawaharlal Nehru National Urban Renewal Mission (JNNURM). The 2015 'Smart City guidelines', envisaged by the Indian state as well as comprehensive and total urban plans like Atal Mission for Rejuvenation and Urban Transformation (AMRUT), makes governance ever more total. There are specific directions issued to states regarding schemes that direct funds to Urban Local Bodies (ULB). Individuals are encouraged to enter Public Private Partnerships (PPPs). Unprecedented centralisation can be seen in comprehensive conditionalities involved. Rupees 203,976 crore allotted to smart cities have to be operationalised through special purpose vehicles (SPVs) modelled on corporate entities and headed by CEOs.

Ever more responsibilities are invested not only in individual state but over local ecologies that have least say in the designs that ensue. The guidelines run complementary in other state schemes like AMRUT or Swachh Bharat Abhiyan (SBA). All these have an investor friendly ecological securitisation that allays environmental and ecological risks in terms of capital. A case in point is the Gujarat International Finance Tec-City (GIFT) technological city. The exceptionalities that operated through urban development corporations in essence diverted pasture land and land deemed 'waste' to upcoming smart space. The securitisation discourse empowers SPVs get ever more to impose more investor friendly designs. The risks and responsibilities are in effect borne by the human-non-human inhabitants and local governments.

Ironically the proliferating risks from land development and real estate are only perceived as inadequate imposition of smart policies which in reality are part of the problem. The new laws that promote corporate private investments as a conditionality on designated regions as well as those that sanctioned foreign direct investments in real estate or the new provisions of speeding up environmental clearances do not come under the purview of disaster reliefs. Disaster reliefs are at best symptomatic in action. The logic in the desire to promote development that sits at odds with the local ecosystems has spilled over urban spaces even to the Himalayas (Bisen, 2023; Panwar, 2024) or Western Ghats (Madhav Gadgil Commission, 2013). In every instance state securitisation literally greenwash the causes out!

Another frame is that of green securitisation which refers mostly to securitisation of financial assets. Accordingly, collaterals are issued on those assets deemed to be green. Such asset creations have been on in US, China, France or even at a much lesser extent in India. Apart from all the questions regarding economisation of ecology, there are structural issues as well. For instance, the proceeds from the loan issued so can be used in polluting sectors (e.g. hydro carbon intensive). This becomes more pertinent in India where some of the leading private corporate groups are leading importers of coal (Adani group) or own refineries (Reliance Industries). And green securitisation can easily liquidate the stated intend, however flawed these schemes may be even otherwise. As with most other climatic regimes, the heterogeneous regulatory frameworks in countries reinforce state sovereignties, more so with those large and powerful states.

At a philosophical level, we have been seeing everyone—from individuals, corporates, companies and states becoming green. In other words, there are entrenched assumptions of naturalisation of ecological perspectives as it happens within all these. Once green securitisation policies are adopted, there is a comfortable numbness in thoughts on what has gone into its making and later assumptions of balanced harmonies. The heterogeneous and complex ecologies in question are or have never been balanced because within each may be placed the variant responses, effects and consequences of planning and policies. What is often forgotten, whether this is China or India or the US.

China for instance has come out big on the so-called clean energy transition technologies, the market of which depends heavily on minerals like lithium, cobalt, zinc and other rare earth materials that go into the panels and turbines. If there is a presumption of harmony that precludes any thought on the environmental and human impacts in those sites where these investments are being made. The green transitions and securitisation directives of China, India, US, EU or Canada cannot be separated from the contested ecologies within or 'overseas', whether they are in the DRC, Peru or Zimbabwe. Technically sound environmental practices are often abstractions of unjust business models that do not limit itself to bare carbon or fossil fuel. Instead, they are linked to structures of particular economic systems that operate as subsets of a grand neoliberal capitalist routines. The accumulations by dispossessions, environmental damages or neocolonial conflicts do not feature in these apolitical narratives. The International NGO, *Global Witness*, for example documents the violence and murders involved in many of these places where mining, agribusiness and other green security investments have gone into.

We can trace the genealogy of the green securitisation agendas to the broad post 'visible cold war' situations wherein the traditional threats, albeit never forgotten, were broadened into foci like ecology and environment (Trombetta, 2008). The constructivist turns and Copenhagen school of course focussed on this whole discourse and how and who handles these. Things become even more complicated because states have transformed their morphologies beyond identifiable forms. Corporate states can operate simultaneously and complementarily as capitalist democracies, communist states and liberal democracies while administering post-political fixes that seek to tweak the economy, urban processes or ecology. Green security solutions are designed to create a general impression that everything is in order. Everything may, in fact, be in order. It is just that the order is sustained by a silence of politics (Swyngedouw, 2018; Zizek, 2018). Carbon-centric fetishisation of ecology exemplifies a post-political process guided by techno-managerial *dispositifs* (Swyngedouw, 2018). This makes the political appear more spectral. Post-political good governance creates the enabling conditions that emerge outside traditional ideas of state, but at the same time guided inwards at every point by the variant state orders. The perceived and projected failures of states are used to reinforce this order ever more. Green securitisation and carbon fetishes (involved in carbon trades, etc.) take the forms of institutional practices nurtured by the state and international bodies like the EU, corporate apparatuses like the World Bank or non-state actors. But there is an accompanying rhetoric of participation, inclusion or empowerment, unlike the rhetoric of entrenched state hierarchies.

The third model of compensatory framework is perhaps the most contentious because it is a proposal by the think tank styled organ called Niti Aayog to create competitive neoliberal ecosystems like the ones in between territories directly under union government (viz. Union Territories). The planned Nicobar infrastructural project that converts more than 160 square kilometres of mostly tropical forest is one such fallout. Again 'state interests or national interest' become exceptional situations under which endemic ecosystems or indigenous lifeworld may be decimated. To make matters worse Compensatory Afforestation regimes get invoked through developmental companies. Non-forested or revenue lands in places far away could first get abstracted and then equated in value according to the compensatory schemes. Even degraded lands could be equated with the forested Nicobar Island. In addition, the provisions like conversion of non-forested lands for paid compensatory plantations also are a centre heavy scheme. Compensatory Afforestation in other words is an ecological monetisation plan. The investors also pay a 'net present value' to forest department. The enaction of laws on the direction of supreme court that formalised and consolidated the previous compensatory infrastructures (ever since the 80s) also created clear channels of monetisation. The Greater Nicobar Island (GNI) project, a composite of urban spaces, power plants (defended as green or solar!), an airport as well as an international transhipment terminal, involves a massive conversion of unique ecosystems with human and non-human relationships. For NITI Aayog what gains precedence is the location of this biome on a strategic sea map, the corporate private interests as well as the steps like the BRI by a Chinese state which itself is pursuing a gamut of green abstractions. An adjunct phrase of 'eco-friendliness' to business, trade or leisure is enough to make every space greenfield.

The clearances and Environmental Impact Assessments (EIAs) are incorporated in a compensatory discourse as if the endemic species like crab eating Macaque or the Shompen people's ecological niche around Galathea Bay could be easily compensated in a faraway region in Haryana. Even the provisions that are there for setting terms of reference for environmental assessment was flaunted (Sekhsaria, 2022) when another private group started fieldwork much ahead in time. Thus, apart from the exception given to land or ecological alienation in the name of 'public purpose', the existing provisions of self-governance in Panchayats

(Extension to Scheduled Areas—PESA [Ministry of Law and Justice (Legislative Department)], 1996) or the nuanced entitlements of the Forest Rights Act (Ministry of Law and Justice (Legislative Department), 2006, are negated. The Compensatory Afforestation Fund Act (CAFA) 2016 and rules that came into operation since 2018 exemplify a structural transformation of governance.

The glimmer of embedded perspectives and local roles in perceptions, planning and procedures of ecologies that Forest Rights Acts or PESA carried are pre-empted in emergent bureaucratisation processes. The think tank infrastructures have created top-down procedures and provisions mediated by 'experts' that in effect feudalise forests and annihilate any remaining provision to address the historical injustice inflicted on tribals, non-human entities, as well as their complex relationships ever since colonial times.

The scenarios and policy contexts involved in, risk and risk-based infrastructures, green securitisation or compensatory frameworks exemplify offer critical understanding of 'greening' within India as well as by implication, the interlinked global political economies. These are only springboards in order to go beyond. In our understandings of the discourses of security that are premised on risk and risk ecologies (societies *pace* Ulrich Beck, 1992), abstractions and alternate securitisations, we still have to put the state into clear relief. It still remains the main deployer of horizontally networked heterogeneous orders that work globally. But if we take divergent situations of India, China or Indonesia, there are several ambiguities built in the existing greening provisions. Ironically, the ambiguities become the means whereby post-politics deal with any crisis in legitimation. The abstract and unclear discourses of eco-friendly, green, stakeholder-based, carbon neutral, expert-based or techno-scientific solutionist are replicable globally and greenwash any heterogeneous or complex ecological relationships. Such procedural abstractions and unclarities are corollary to neoliberal capitalism. On the one hand, ever since the nineties, a crisis of developmental state in India (welfare states and revolutionary states elsewhere) or that of Keynesian obligations built into capitalisms has legitimated the 'withdrawal of state' rhetoric. The state though has never gone away. As the contemporary green securitisations suggest, it has been redeployed in a post-political environ where ecological solutions are seldom politicised.

In all these scenarios, we have moved beyond but well through the systematic and technical address of hazards that Beck (1992) warned about with respect to modernisation processes. Then it was about how hazards generated by the system led to futuristic risk-modulated orders. Now it is about a more entrenched techno-solutionist post-political environ. It suggests an incapacitated state where only think tanks and experts work. But any possible assumption of the loss in regulatory capacities is misplaced. It is just that the regulatory role has transformed to that of managerialisms and responsibilities have shifted on to the actors that are most affected and, on the ground (whether that is the leatherback turtle, the Crab Eating Macaque, the mangroves buffered the impact of tsunamis in 2004 or the indigenous communities). All of these have to be resilient and thereby get celebrated. They will have to be the sole beasts of burden in sustainable orders proposed.

CONCLUSION

From the vignettes and frames as well as the premises in which securitisation theories as well as the constructivist schools become significant, we have to move beyond into the ritually codified policymaking domains taking shape in heterogeneous contexts globally. We need to listen to voices from both ecologically and socially complex terrains but at the same time not falling into post-political frames. If I may draw from Zizek of Badiou, the naturalisation and ecologisation of politics are the new fetish. There is an assumed consensus politics, in post-political suggestions. But greening, as it happens in India as well as the symbiotic state domains elsewhere, demonstrates the regulatory state getting ever more entrenched, managerial and pastoral in policies and prescriptions.

The risk ecologies that green securitisation as a discourse draws upon as well as the techno-managerial solutionism in compensatory schemes hit the limits soon as the local social worlds or ecological relationships are brought in. With all the possibilities of dwelling on the interpellations and the authorship built into securitisation frames, there still remains a need to dwell on how ecology can be used as the new opium of the masses. We may take here the post-disaster risk-aversion measures, fetishisation of carbon in climate change, green developmentalism, etc. In every case, nature is co-constituted in an unwillingness to politicise ecology. There is a failure to see that there is no foundational assumption in nature; rather nature and ecology are replete with entrenched co-evolutions and inter- and intra-species hybridities. The way ahead is to go further from the possibilities of constructivist or securitisation theories into nuanced 'recognition' of unpredictability, radical contingency, non-linearity or variabilities in candidate ecologies. This can ward off managerial visions and fixes. This cannot be just about a maverick flood, a lack of smart solution, development or alien greenhouse gases. It is about systemic disasters or their regularities, the problem of smart solutions or cities, land alienations, repressions, speculative trading, monetised abstractions as well as neocolonial controls. After such radical recognitions, greening of ecological solutions can no more be what it was.

References

Beck, U. (1992). *Risk society: Towards a new modernity* (M. Ritter, Trans.). Sage Publications.

Bisen, A. (2023, February 23). *The elusive smartness in India's smart cities*. The Wire. https://thewire.in/government/the-elusive-smartness-in-indias-smart-cities

Booth, K. (1991). Security and emancipation. *Review of International Relations, 17*, 313–326.

Floyd, R. (2007). Towards a consequentialist evaluation of security: Bringing together the Copenhagen and Welsh Schools of security studies. *Review of International Studies, 33*, 335.

Madhav Gadgil Commission. (2013). *Report of the Western Ghats, ecology expert panel*. The Ministry of Environment and Forests, Government of India. https://ruralindiaonline.org/en/library/resource/report-of-the-western-ghats-ecology-expert-panel/

Panwar, T. (2024, February 07). *Bleak urban outlay, neglected Himalayan states: Decoding the 2024 Budget's Election Rhetoric*. The Wire. https://thewire.in/urban/bleak-urban-outlay-neglected-himalayan-states-decoding-the-2024-budgets-election-rhetoric

Sekhsaria, P. (2022, November 15). Despite red flags, mega project on ecologically-sensitive Great Nicobar Island gets green signal. *Scroll.in.* https://scroll.in/article/1037433/despite-red-flags-mega-project-on-ecologically-sensitive-great-nicobar-island-gets-green-signal

Swyngedouw, E. (2018). *Promises of the political: Insurgent cities in a post-political environment*. MIT Press.

Trombetta, M. J. (2008). Environmental security and climate change: Analysing the discourse. *Cambridge Review of International Affairs, 21*(4), 585–602.

Wyn Jones, R. (1996). "Travel without maps": Thinking about security after the cold war. In J. Davis (Ed.), *Security issues in the post-cold war world* (pp. 196–218). Edward Elgar.

Zizek, S. (2018, December 3). If we want to survive on this planet, we need to abandon the cause of the nation state. *New Statesman*. https://www.newsta tesman.com/politics/2018/12/if-we-want-survive-planet-we-need-abandon-cause-nation-state

Role of Epistemic Communities in Shaping India's Environmental Policies: A Constructivist Perspective

Chandran Komath

Abstract This paper employs insights from the Constructivist Theory of International Relations to examine the impact of 'epistemic communities' in shaping India's environmental policies, particularly regarding climate change, and to analyse India's emergence as a responsible environmental actor on the global stage. It argues that India's proactive engagement with diverse global environmental epistemic communities, such as the Intergovernmental Panel on Climate Change, the Convention of Parties, International Solar Alliance, and the Convention on Biological Diversity, over the past three decades has significantly influenced the integration of environmental considerations into India's national identity ('greening the nation') within the global environmental discourses. In conclusion, the paper claims that India's collaboration with international epistemic

C. Komath (✉)
Department of Political Science, Government College Kottayam, Kottayam, Kerala, India
e-mail: komathchandran@gmail.com

© The Author(s), under exclusive license to Springer Nature Singapore Pte Ltd. 2024
J. T. Karackattu et al. (eds.), *Environmental Securitisation in India and China*, https://doi.org/10.1007/978-981-97-9160-6_4

communities enhances its credibility as a responsible actor in the context of global climate policies and mitigation strategies.

Keywords Constructivism · Epistemic communities · Climate change · Environmental policy

INTRODUCTION

This chapter draws on insights from Constructivist Theory in International Relations to explore the contributions of 'epistemic communities' in India's engagements with global environmental question especially within the framework of climate change. It is argued that India's environmental policies are deeply driven by international environmental discourses including ideas, norms, principles, and identities which significantly shape its interactions with the international community (Biermann & Pattberg, 2012; Dubash, 2012; Harris, 2022). To justify this perspective, it is essential to first explore the innovative dimensions of Constructivist Theory by tracing its origin, key concepts, and contributions from prominent scholars who have significantly advanced this theoretical framework.

In the late twentieth century, Social Constructivism emerged as a critique of, and response to, the limitations of positivist political theories in explaining and understanding global politics and it gained prominence in the 1990s, largely through the writings of Nicholas Onuf, Alexander Wendt, and Martha Finnemore (Finnemore, 1996; Kratochwil, 1991; Onuf, 1989; Wendt, 1999). Unlike neorealism and neoliberalism, which largely prioritise material power and state preferences in the so-called anarchical conditions, Constructivism underscores the centrality of ideas, concepts, values, perspectives, and identities in influencing the structure, dynamics, and outcomes of the international system. At its core, Constructivism emphasises the importance of socially constructed realities in influencing and reshaping international relations. Therefore, in Constructivism, we find the beginning of a social theory of international relations (Wendt, 1999). At the outset, an overview of these foundational concepts and ideas underpinning Constructivist Theory is outlined here.

One key aspect of Constructivism is its focus on the importance of intersubjective meanings (Wendt, 1999, p. 160). In Constructivist

Theory, meanings are not seen inherent in objects or actions but are socially constructed through continuous processes of interaction and interpretation. This perspective enables a more nuanced understanding of phenomena such as state behaviour, cooperation, and conflict resolution in international politics, which cannot be fully explained by material factors alone. Constructivists contend that state and non-state actors in international politics source their identities, interests, and preferences based on intersubjective interactions than from objective material conditions (Katzenstein, 1996; Wendt, 1999). This idea challenges the realist/neo-realist notion of a fixed and anarchic international system, proposing instead that the meanings attributed to concepts like power, security, and cooperation are shaped by many forms of social interactions. Thus, Constructivism posits that the societal spheres function as a product of human beliefs, meanings, and countless forms of interaction, with multiple players deriving their individualities and belongings influenced by collective contexts than inherent characteristics.

Ideas and norms are crucial in shaping state behaviour, as constructivists assert. Shared beliefs and norms significantly influence state actions, driving the formation of alliances, policy adoption, and acceptance of international institutions (Finnemore, 1996). For instance, the global diffusion of democratic norms and principles often encourages states to prioritize human rights and democratic governance in their foreign policies. Constructivism also underscores the role of ideational and identity-related factors in determining state behaviour in the international system. States construct their identities in relation to others, drawing on shared cultural, historical, and ideological factors. These factors, in turn, shape how states perceive their national interests and engage with other actors on the international stage (Finnemore, 1996).

Furthermore, Constructivism emphasises the dynamic nature of international politics, highlighting how identities and interests evolve over time through processes of learning and socialisation. As Wendt observes, "Realists of all stripes believe that states do what they do because it is in their national interest, and that the national interest is self-regarding with respect to security" (Wendt, 1999, p. 113). In contrast to the realist portrayal of eternal anarchy and power politics, Constructivism recognises the transformative potential of new ideas and norms (Wendt, 1992). States, as socially embedded entities, have the capacity to reinterpret and reshape their self-identity and security via interactive process and social knowledge. According to Finnemore, states are "embedded in dense networks of transnational and international social relations" (Finnemore, 1996, p. 2), which opens avenues for cooperation and conflict resolution in various ways.

In short, distinct from the positivist theories of international politics, Constructivist Theory offers a rich and nuanced perspective on existing and emerging forms of global interactions by stressing the centrality of concepts, rules, and characteristics in forming state practices and dynamics of world politics (Kratochwil, 1991). By challenging traditional international political theories' ontological and epistemological assumptions concerning states and systems, Constructivism provides a socially oriented analytical framework for understanding the complexities of historical and contemporary international systems, structures, and processes (Wendt, 1999).

NORMS, IDEAS, AND IDENTITY IN INDIA'S ENVIRONMENTAL POLICY

Constructivism underscores the essential part of *epistemic communities*—knowledge-centred networks of experts and professionals—in shaping state preferences in connection with various questions in global politics (Haas, 1992). These knowledge-based societies are playing an active role in the contemporary international system. They are also central to generating and disseminating scientific and policy-related knowledge concerning various pressing global issues (Haas, 2016). Haas also emphasises the importance of 'usable knowledge' that speaks truth to power. For Haas "usable knowledge encompasses a substantive core that makes it usable for policy-makers, and a procedural dimension that provides a mechanism for transmitting knowledge from the scientific community to the policy world and provides for agency when theorising about broader patterns of social learning, policy-making, and international relations" (Haas, 2004a, p. 573).

Against this background, India's highly interactive and dynamic process on global environmental issues can be examined at various levels. As previously discussed, norms play a crucial role in shaping state behaviour, with norm diffusion and adoption becoming central to debates on contemporary global environmental governance (Bernstein, 2013; Biermann & Pattberg, 2008; Finnemore & Sikkink, 1998; Harris, 2022; Hoffmann, 2013). India's active participation in numerous international environmental agreements demonstrates the diffusion and adoption of global norms related to sustainable development and climate action, as reflected in its national plans and actions (Dubash, 2013, 2019). It is possible to observe that the evolving global discourse on environmental responsibility and notions of global environmental governance clearly influence India's engagements related to climate questions.

On the one side, India's proactive engagement with global environmental collectives contributes to the construction of its national identity as a responsible global actor, committed to addressing shared environmental challenges of the contemporary era. On the other side, India's involvement in various global climate forums serves as a platform for socialisation, where it learns and internalises environmental norms through interactions with other states and organisations. India's engagement with international scientific and environmental communities has facilitated the adoption of best practices and the integration of scientific knowledge into policy decisions. As mentioned earlier, Constructivist Theory put forward by scholars like Haas emphasises policy learning and adaptation as critical aspects of contemporary global climate discourse (Haas, 2016). It could be argued that India's responsiveness to global norms in areas like renewable energy, sustainable development, and climate crisis management illustrates a process of policy learning, whereby strategies are adapted based on international experiences and evolving ideas.

Constructivist IR Theory further emphasises the functions of concepts, rules, ideals, and other discursive practices in determining the conduct of states (Kratochwil, 1991). India's participation in international forums and commitment to global environmental goals contribute to various discursive practices that underscore the significance of viable progress and effective environmental management. Identity contestations have a key influence in the development and negotiation of national identities within contemporary global politics (Onuf, 1989). In India's case, negotiating its identity as a responsible environmental actor reflects a balancing act between global expectations and domestic constraints, such as developmental priorities and resource needs. This dynamic process illustrates the concept of 'constructed preferences', where India aligns global environmental norms with its domestic developmental imperatives (MEFCC, 2022). From a constructivist perspective, it is possible to conclude that India's engagement with global environmental epistemic communities represents a dynamic process of identity construction, socialisation, and policy adaptation. This process is shaped by evolving ideas and interactions around global warming and climate mitigation strategies. There may be other social and political factors which influence India's interactions in the case of global environmental discourses; however, this chapter is primarily concerned with explaining the working of epistemic communities and illustrating how India presents itself as an accountable and credible environmental player in global affairs.

Idea of Epistemic Communities in Constructivist Theory

The idea of "epistemic communities" is an innovative concept theorised within the constructivist understanding and explanations on global politics, notably by Peter M. Haas(Haas, 2016). According to Haas "An epistemic community is a network of professionals with recognized expertise and competence in a particular domain and an authoritative claim to policy-relevant knowledge within that domain or issue-area" (Haas, 1992, p. 3). These communities wield considerable influence in shaping state behaviour and policy outcomes by guiding how policymakers perceive, interpret, and address complex global challenges. Haas also identified four defining characteristics peculiar to epistemic communities. These are "(1) a shared set of normative and principled beliefs, which provide a value-based rationale for the social action of community members; (2) shared causal beliefs, which are derived from their analysis of practices leading or contributing to a central set of problems in their domain and which then serve as the basis for elucidating the multiple linkages between possible policy actions and desired outcomes; (3) shared notions of validity- that is, intersubjective, internally defined criteria for weighing and validating knowledge in the domain of their expertise; and (4) a common policy enterprise-that is, a set of common practices associated with a set of problems to which their professional competence is directed, presumably out of the conviction that human welfare will be enhanced as a consequence" (Haas, 1992, p. 3).

Various epistemic communities exert their influences in different ways. Haas further notes that "Members of transnational epistemic communities can influence state interests either by directly identifying them for decision makers or by illuminating the salient dimensions of an issue from which the decision makers may then deduce their interests. The decision makers in one state may, in turn, influence the interests and behavior of other states, thereby increasing the likelihood of convergent state behavior and international policy coordination, informed by the causal beliefs and policy preferences of the epistemic community" (Haas, 1992, p. 4). A brief examination of various epistemic communities working in the field of human rights, climate issues, health, and disarmament will help to highlights their significant role in shaping global governance through expertise, shared knowledge, and transnational collaboration. For instance, in global health governance, epistemic communities comprising public health experts, NGOs, and international organisations have contributed to policies addressing infectious diseases, maternal health, and healthcare access in emerging economies (Dalglish et al., 2015). Similarly, inside the domain of arms control and non-proliferation, communities of nuclear scientists, diplomats, and disarmament advocates have played a pivotal role in developing disarmament treaties, verification mechanisms, and diplomatic negotiations aimed at curbing the proliferation of nuclear weapons (Adler, 1992; Lane, 2010). In global climate policy, epistemic communities of climate scientists, policymakers, and social activists have been instrumental in shaping international climate agreements, providing scientific expertise, building consensus, and mobilising public support for mitigation and adaptation measures (Haas, 2004b).

Based on the above discussions, four key dimensions can be identified in the conceptualisation and operational aspects of environmental epistemic communities. The first dimension pertains to the idea and practice of 'shared knowledge and expertise' about environmental governance and specific policy issues (Haas, 2016; Montana, 2021). Environmental epistemic communities often share understanding of particular issue areas, grounded in specialised knowledge and expertise. These communities comprise diverse members, including academics, scientists, bureaucrats, and practitioners, who converge around common norms, beliefs, and epistemic frameworks that shape their analyses and approaches to problem-solving. The second dimension focusses on 'norm entrepreneurship' (Haden & Seybert, 2016; Zwolski & Kaunert, 2011). As norm entrepreneurs, environmental epistemic communities play a critical role in advocating for specific policy solutions and promoting their adoption at national and international levels. Leveraging robust networks of communication and collaboration, these communities influence policy agendas, shape public discourse, and build broad support for their preferred strategies to address global environmental challenges.

The third dimension of environmental epistemic community centres around 'policy learning and innovation' (Dunlop et al., 2024; Haas, 2004a). These communities foster policy learning and innovation by providing policymakers with new ideas, evidence-based analyses, and best practices from their respective fields. Through interactions with experts and practitioners, policymakers gain insights into complex issues, explore alternative policy options, and adapt strategies in response to evolving circumstances (Wesselink et al., 2013). The fourth dimension concerns the function of 'transnational advocacy networks' in global affairs (Park, 2023). Often functioning as transnational advocacy networks, environmental epistemic communities collaborate across national borders to address shared concerns and exert influence over global governance mechanisms. By forming alliances with like-minded actors, these communities amplify their impact, enhance their visibility, and mobilise resources to advance their policy agendas on the international stage (Park, 2013). This perspective offers several insights about the changing dynamics of contemporary international relations and also mechanisms driving global change in response to climate-related challenges.

FORMATION OF GLOBAL ENVIRONMENTAL EPISTEMIC COMMUNITIES

The formation of international environmental epistemic communities, particularly in connection with atmospheric changes and other broader ecological challenges, underscores the significant role of experts and professionals in shaping international policy responses (Adler & Haas, 1992). These epistemic communities, which include scientists, policymakers, activists, and other stakeholders, are united through the mutual comprehension of environmental challenges, standards, and potential solutions. Peter M. Hass observes that "Epistemic communities are one possible provider of this sort of information and advice. As demands for such information arise, networks or communities of specialists capable of producing and providing the information emerge and proliferate. The members of a prevailing community become strong actors at the national and transnational level as decision makers solicit their information and delegate responsibility to them" (Haas, 1992, p. 4). The cooperative efforts of them transcend national boundaries and also made a significant contribution to global initiatives aimed at addressing climate change and other environmental crises (Haas, 2004b).

One prominent example of a global environmental epistemic community is the International Panel on Climate Change (IPCC) (Hoppe, 2010). The IPCC was co-created by the United Nations Environment Programme (UNEP) and the World Meteorological Organization (WMO). It unites thousands of scientific researchers and experts from across the globe to study and appraise the important scientific investigations on global warming and other climate issues. Through its comprehensive reports, the IPCC provides policymakers with authoritative information and recommendations, significantly influencing climate change negotiations, policy decisions, and global public awareness efforts. The Conference of the Parties (COP), founded in 1992, works as the topmost policy-formulation body of the UNFCCC. COP meetings convene representatives from nearly 200 countries to negotiate and implement international actions to battle climate change and related issues. These annual gatherings provide a vital platform for nations to negotiate and establish strategies focussed on lowering carbon emissions, understanding the consequences of climate change and delivering economic and technical assistance to vulnerable countries.

Another key example is the Climate Action Network (CAN), an alliance of non-governmental organisations (NGOs) dedicated to promoting ambitious climate action. CAN functions as a platform for coordination and collaboration among environmental activists, experts, and advocates, facilitating the transfer of data, finest practices, and policy strategies (Newell, 2000). Through mobilising public support and lobbying policymakers, CAN and its members have been instrumental in shaping global climate accords, including the Paris Agreement. Moreover, academic institutions, research organisations, and think tanks contribute to global environmental epistemic communities through their research, publications, and advocacy efforts. These institutions provide vital scientific expertise, policy analysis, and empowerment support for governments, NGOs, and additional participants engaged in environmental governance and climate action (Biermann & Pattberg, 2012). Generally, we can say that, the formation of global environmental epistemic communities highlights the importance of knowledge-sharing, collaboration, and collective action in addressing pressing environmental challenges, particularly climate change. By leveraging the expertise and insights of diverse stakeholders, these communities have a critical part in shaping effective policies, strategies, social actions, and solutions to protect the earth for future generations.

Trajectory of India's Engagement with Global Environmental Epistemic Communities

India's engagement with global environmental epistemic communities has evolved significantly over time, reflecting the nation's increasing awareness of the significance of environmental issues as well as its desire to take an active part in global environmental governance (Atteridge et al., 2012; Dubash, 2012, 2019; Youdon & Bajaj, 2022). In the following section, I will provide a brief assessment of India's interactions with global epistemic communities and highlight significant achievements in this regard.

India has been actively involved in international climate negotiations, championing the concept of shared but differentiated obligations and advocating the necessity for industrialised nations to assume a larger role in tackling climate change as they are the primary contributors to carbon emissions in the past and possess the money and technology to aid emission reduction and climate adjustment in vulnerable regions (Gupta et al., 2015; Page, 2013). India viewed Paris Agreement in a constructive manner and has agreed to ambitious goals for lowering carbon emission (Prasad & Kochher, 2009; Sengupta, 2019). In addition to climate action, India is home to rich biodiversity and has undertaken various initiatives to conserve and protect its natural heritage. India has ratified global treaties like the CBD and enthusiastically contributed in global efforts to protection of biodiversity conservation and the responsible use of environmental assets (Goyal & Arora, 2009).

India's engagement with global environmental epistemic communities has its beginning in the early stages of international environmental diplomacy (Dubash, 2019; Najam et al., 2006). Initially, India adopted a cautious approach, prioritising national development goals over environmental concerns (Gupta et al., 2015). However, over time, India has increasingly acknowledged the urgent need to tackle ecological problems like climate variation and unsustainability of ongoing developmental works. India's engagement with global environmental epistemic communities is characterised by cooperation, negotiation, and advocacy (Vihma, 2010). While the country strives to safeguard its national interests and sovereignty, it also acknowledges the necessity of collective action and multilateral cooperation to tackle shared environmental challenges. India plays a dynamic part in various international settings like the UNFCCC, in which the country cooperates with the help of scientific experts, policymakers, and civil society organisations to shape global environmental policies and initiatives (Dasgupta, 2019).

Over the last few years, India has assumed a growing role in Conference of the Parties (COP) meetings, contributing to negotiations as well as advocating for its interests in climate policy. Indian delegates engage in various working groups and committees within the COP framework, offering expertise and insights on issues such as mitigation, adaptation, finance, and technology transfer (MEFCC, 2023). India's proactive engagement underscores its commitment to sustainable development and its realisation of the importance of collaborative action in combating change. India has achieved several notable accomplishments through its engagement in COP meetings. These include playing a constructive role in the negotiation of key accords like Paris Agreement in 2015, where India advocated for the concept of fairness and but unequal obligations (Parikh & Parikh, 2016). India's pledge to the Paris Agreement, notably its promise to increase the proportion of sustainable and non-carbon-based alternative energy sources in its overall energy pool, demonstrates its willingness to engage constructively with global environmental epistemic communities (Government of India, 2022; Lakshmanan et al., 2017). Additionally, India has implemented ambitious domestic climate strategies and actions plans like the National Action Plan on Climate Change (NAPCC) and took a leading part in founding the solar alliance, demonstrating its dedication to combating environmental problems.

India actively engages with the IPCC by contributing scientific expertise, data, and research findings to the panel's assessments. Indian scientists and researchers participate in IPCC working groups, contributing to the assessment of climate science, impacts, and mitigation options (Sengupta, 2019). India's engagement with the IPCC reflects its resolve to confront climate issues and its realization of the value of scientific and technical collaboration in informing climate policy. India's achievements in engaging with the IPCC include its contributions to the panel's assessments and its initiatives to advance climate actions at the national and international levels (Gupta & Ghosh, 2023). By actively participating in the IPCC process, India demonstrates its commitment to evidence-based policymaking and its willingness to cooperate with various international collectives to confront the pressing posed by climate change.

India made several significant achievements in its engagement with global environmental epistemic communities. These include initiatives in renewable energy development, climate policy advocacy, biodiversity conservation, and more (Ganeshan & Bhattacharjya, 2022). India has become a leading force in the advancement of sustainable energy production on the global stage with its ambitious targets for expanding its solar, wind, and hydroelectric power capacity. Through laudable programmes like the International Solar Alliance (ISA), India has collaborated with other countries to encourage solar energy growth and sharing of solar technology. The ISA aims to boost solar energy expansion and enhance cooperation among countries with abundant solar resources. India's leadership in the ISA has helped mobilise international support for solar energy development, with over 100 countries joining the alliance (Mishra, 2024).

The ISA seeks to address challenges related to energy access, energy crisis, energy sustainability, and other environmental issues by harnessing the abundant solar resources available among nations. India's engagement in the ISA reflects its commitment to renewable energy development and its recognition of the potential and capacity of solar energy to fulfil increasing energy demands in the context of national development and progress. India, as a formative member of the ISA and monetary support to advance the basic and evolving objectives of the alliance. Experts from India and other member countries contribute knowledge, experience, and best practices to ISA initiatives and working groups. They collaborate on projects related to solar energy research and development, technology transfer, capacity-building, and policy coordination. Indian experts play a key role in sharing lessons learned from India's own experiences in scaling up solar energy deployment and overcoming barriers to adoption (Jha, 2023).

The ISA has achieved several significant milestones since its inception, including the launch of various programmes and initiatives to promote solar energy deployment. These include the establishment of the World Solar Bank to mobilise funding for solar projects, facilitate technology transfer, and encourage solar power as a remedy to global warming (World Bank Group, 2016). The ISA represents a unique base for partnership among countries with abundant sun light to speed up the shift to clean and sustainable energy systems, acting a crucial part in worldwide initiatives to combat ecological issues and attain national development goals.

The Convention on Biological Diversity (CBD) is an agreement focussed on the protection and preservation of biodiversity and ecosystems. The CBD aims to tackle the continuous decline of biodiversity across the globe while recognising its vitality for the maintenance of ecosystem stability and sustainability of species life. India actively engages in the CBD process, participating in meetings, negotiations, and initiatives aimed at implementing the objectives of the convention. Indian representatives contribute expertise and perspectives on biodiversity conservation, sustainable development, and traditional knowledge systems (Lanzerath & Friele, 2016; Tandon et al., 2017). India's engagement reflects its rich biodiversity and its commitment to protecting and utilising natural resources in a sustainable manner. Indian experts, including scientists, policymakers, and representatives from indigenous and local communities are essential in shaping discussions and decisions concerning our ecosystems. They contribute scientific research, traditional knowledge, and policy recommendations to CBD meetings and working groups, informing efforts to address biodiversity loss, habitat degradation, and species extinction (MEFCC, 2019).

India has achieved several notable accomplishments through its engagement in the CBD process. The Nagoya Protocol on Access and Benefit-Sharing is an example in this regard. The basic objective of this agreement is to guarantee the just and balanced distribution of advantages gained from utilizing genetic resources. India has also implemented various national-level initiatives and policies to preserve biodiversity, safeguard endangered species, and encourage sustainable practices in land and resource management. India's engagement in the CBD reflects its commitment to biodiversity preservation and sustainable progress along with its recognition of preserving natural ecosystems for future generations (MEFCC, 2019).

IMPACTS OF INDIA'S ENGAGEMENT WITH GLOBAL ENVIRONMENTAL EPISTEMIC COMMUNITIES

Based on the above discussions, it is possible to say that India has accumulated and developed a significant scientific knowledge base and expertise through active participation in various global epistemic communities, such as the IPCC, COP, CBD, and ISA among others. India has also tried to engage with these platforms in a very constructive manner, thereby gained access to cutting-edge scientific research, best practices, and policy insights, all of which have been instrumental in shaping its domestic climate change and mitigation strategies.

Participation in the IPCC has provided Indian scientists and experts with the opportunity to support analysis and evaluation of global environmental change, its impacts, and so on. Through involvement in IPCC working groups and research activities, Indian researchers have gained a deeper understanding of climate change dynamics, regional vulnerabilities, and adaptation measures. This enhanced scientific knowledge has been instrumental in shaping India's climate policies and approaches for lowering carbon emissions and adjusting to the effects of environmental changes (Sengupta, 2019).

India's participation in COP meetings has enabled the country to actively contribute to global climate negotiations and agreements. By participating in COP discussions, India continually endorsed the position of 'common but differentiated responsibilities' and emphasised the requirements like monetary and technical provisions for emerging economies in the developing world (Gupta et al., 2015). India's contributions to COP decisions have influenced the creation of international climate frameworks and commitments, which, in turn, inform the development of India's own climate policies and mitigation strategies.

Participation in the CBD has facilitated India's collaboration with other countries on biodiversity conservation efforts and sustainable resource management practices. By sharing scientific knowledge, traditional wisdom, and best practices with the global community, India has contributed to the creation of international biodiversity policies and initiatives. This engagement has also shaped India's national strategies for conserving biodiversity, protecting ecosystems, and promoting sustainable development.

Involvement in the ISA has fostered collaboration among solar-rich nations to promote solar energy deployment and technology transfer. By sharing expertise, research findings, and policy experiences with other ISA members, India has enhanced its capacity to harness solar energy resources and integrate renewable energy into its national energy mix. This collaboration has informed India's efforts to develop and implement policies and programmes aimed at expanding solar energy infrastructure, reducing carbon emissions, and transitioning to a green and sustainable economic power.

In conclusion, India's participation in global epistemic communities has significantly enriched its scientific knowledge base, enhanced its technical capabilities, and influenced the development of its climate policies and mitigation strategies. By engaging with international platforms like the IPCC, COP, CBD, and ISA, India has contributed to global initiatives to fight climate problems, biodiversity loss, and balanced growth and progress of less developed countries, while advancing its own national priorities. Additionally, India has already incorporated several Sustainable Development Goals (SDGs), including environmental goals, into its policy framework. The country is also actively engaged in global efforts to protect endangered species and ecosystems, and global pollution control norms have shaped India's policies on air, water, and soil pollution. Measures to reduce emissions and improve waste management further reflect India's commitment to international environmental standards.

'Greening' the National Identity and Global Environmental Policy

In the past few years, many developing countries, including India, have increasingly sought to 'green' their national identities, integrating environmental considerations into their nation-branding efforts. This trend aims to project an image of environmental stewardship and sustainability (Dinnie, 2008; Duit et al., 2016; Eckersley, 2021; Fan, 2010). Nation branding, which has gained prominence in the era of globalisation, is essential for countries seeking visibility and relevance on the world stage. India, with its rich natural heritage and rising environmental consciousness, has made significant strides in branding itself as a 'green nation' by emphasizing its commitment to environmental conservation and sustainable development (Government of India, 2022). Nation branding in the environmental context influences global perceptions of a country's approach to environmental policies and practices.

Why is 'greening' the national identity significant in contemporary global politics? What role do environmental considerations play in shaping India's national identity? How far has India gone in branding itself as a 'green nation'? To answer these questions, it is essential to first grasp the idea of 'nation branding'. 'Nation branding' generally denotes the purposeful moulding and promotion of a distinctive identity or character of a nation on the international arena (Dinnie, 2008). It involves strategically highlighting a nation's values, culture, and achievements to enhance its global standing, attract foreign investment, increase tourism, and strengthen diplomatic ties. Effective nation branding hinges on aligning a country's policies and practices with the image it seeks to project, emphasizing qualities like environmental sustainability, innovation, and cultural heritage. Simon Anholt, who first popularised the concept, observes that a nation's brand is a dynamic force that shapes global perceptions and impacts its international influence. To bring more clarity to the idea of nation branding, Anholt later introduced the concept of 'competitive identity' of nations, which provides a sound framework for assessing a country's strengths and weaknesses in this area while also allowing for the possibility of reshaping behaviours and actions to improve a nation's global standing (Anholt, 2007). Studies have shown that countries with strong environmental branding attract investment, tourism, and international cooperation, while also fostering national pride and social cohesion (Salah & Mahrous, 2019).

'Greening' the national identity holds particular relevance in the current global context due to several factors. With the world confronting unparalleled ecological issues such as pollution, global warming, and species loss, greening a nation's identity signals recognition of these issues and a commitment to addressing them. This is increasingly significant within the framework of international governance and global rule of law (Koch, 2024). Furthermore, environmental problems are inherently transnational and demand international cooperation for effective solutions. By positioning itself as a green nation, a country can enhance its credibility in global environmental negotiations and foster cooperation with other nations and international organisations.

Greening the national identity also enhances a country's soft power and global influence (Aşkar Karakır, 2018; Nye, 2004). Nations that lead in environmental stewardship are perceived as credible and responsible, which attracts investment, tourism, and diplomatic partnerships. This, in turn, enhances their geopolitical prestige. Moreover, environmental issues resonate deeply with citizens worldwide. Countries prioritising sustainability can strengthen social cohesion and foster national pride, uniting citizens behind government policies aimed at addressing environmental challenges. The green economy transition also offers economic opportunities, including job creation, innovation, and sustainable development (UNEP, 2018). By branding itself as a green leader, a country can attract investments in clean technologies, renewable energy, and green infrastructure, boosting its economic growth and global competitiveness (Barbier,

2010). Additionally, it opens the door for global leadership and responsibility, especially as consequences of climate variations and environmental degradation turn increasingly more severe.

In short, greening a nation's identity is significant in contemporary global politics because it deals with pressing ecological issues while it also enhances a nation's reputation, soft power, and influence in global affairs. It fosters social cohesion, drives economic opportunities, and demonstrates leadership on the global stage (Nitza-Makowska et al., 2024). For India, a prominent nation in the Global South, integrating environmental considerations into its national identity is reflected in policies that emphasise environmental protection and sustainability. India's National Action Plan on Climate Change lays out several imaginative strategies and action plans to assess the impacts of atmospheric changes and promoting sustainable development across various sectors in the national economy (Government of India, 2008).

The NAPCC includes eight key missions that address climate change and promote sustainable growth. These missions focus on enhancing solar energy capacity, improving energy efficiency, promoting sustainable urban planning and transportation, optimising water use, conserving the Himalayan ecosystem, enhancing carbon sinks and biodiversity, supporting climate-resilient agricultural practices, and building capacity for climate change research (Government of India, 2008). These efforts are complemented by initiatives such as the Clean India Mission (Swachh Bharat Abhiyan), which tackles waste management and sanitation. Furthermore, India highlights its environmental initiatives globally through its engagement in emerging global platforms and contracts. Through its endorsement, the Paris Agreement, and proactive participation in COP negotiations, India demonstrates its commitment to global environmental cooperation and leadership (Atteridge et al., 2012). In conclusion, India's efforts to integrate environmental considerations into its national identity reflect a broader trend towards greening national identities, driven by the need for global environmental cooperation and the recognition of sustainability as a key pillar of future development.

Conclusion

India's credibility as a responsible actor in international environmental issues and strategies is underscored via its active engagement with global epistemic communities. The country has proven its resolve in tackling climate issues through its active participation in key epistemic platforms. By contributing scientific expertise, research findings, and policy insights, India has already shown a constructive influence in forming and reforming global climate policies and initiatives. Moreover, involvement of India in various international forums reflects its willingness to collaborate with the global community, exchange knowledge and best practices, and contribute to collective efforts in tackling climate change. This active engagement strengthens India's credibility as a responsible actor in the international climate arena, highlighting its commitment to environmental stewardship and sustainable development on the global stage.

References

Adler, E. (1992). The emergence of cooperation: National epistemic communities and the international evolution of the idea of nuclear arms control. *International Organization, 46*(1), 101–145. https://doi.org/10.1017/S00 20818300001466

Adler, E., & Haas, P. M. (1992). Conclusion: Epistemic communities, world order, and the creation of a reflective research program. *International Organization, 46*(1), 367–390.

Anholt, S. (2007). *Competitive identity: The new brand management for nations, cities and regions.* Palgrave Macmillan.

Aşkar Karakır, İ. (2018). Environmental foreign policy as a soft power instrument: Cases of China and India. *Journal of Contemporary Eastern Asia, 17*(1), 5–26. https://doi.org/10.17477/jcea.2018.17.1.005

Atteridge, A., Shrivastava, M. K., Pahuja, N., & Upadhyay, H. (2012). Climate policy in India: What shapes international, national, and state policy? *Ambio, 41*(1), 68–77. https://doi.org/10.1007/s13280-011-0242-5

Barbier, E. B. (2010). *A global green new deal: Rethinking the economic recovery.* Cambridge University Press.

Bernstein, S. (2013). Global environmental norms. In R. Falkner (Ed.), *The handbook of global climate and environment policy* (pp. 127–145). Wiley. https://doi.org/10.1002/9781118326213.ch8

Biermann, F., & Pattberg, P. (2008). Global environmental governance: Taking stock, moving forward. *Annual Review of Environment and Resources, 33*(1), 277–294. https://doi.org/10.1146/annurev.environ.33.050707.085733

Biermann, F., & Pattberg, P. (2012). *Global environmental governance reconsidered.* MIT Press.

Dalglish, S. L., George, A., Shearer, J. C., & Bennett, S. (2015). Epistemic communities in global health and the development of child survival policy:

A case study of iCCM. *Health Policy and Planning, 30*(Suppl 2), ii12–ii25. https://doi.org/10.1093/heapol/czv043

Dasgupta, C. (2019). Present at the creation: The making of the framework convention on climate change. In N. K. Dubash (Ed.), *India in a warming world: Integrating climate change and development* (pp. 114–141). Oxford University Press.

Dinnie, K. (2008). *Nation branding: Concepts, issues, practice.* Elsevier.

Dubash, N. K. (2012). *Handbook of climate change and India: Development, politics, and governance.* Routledge.

Dubash, N. K. (2013). The politics of climate change in India: Narratives of equity and co-benefits. *Wiley Interdisciplinary Reviews: Climate Change, 4*(3), 191–201. https://doi.org/10.1002/wcc.210

Dubash, N. K. (Ed.). (2019). *India in a warming world: Integrating climate change and development.* Oxford University Press.

Duit, A., Feindt, P. H., & Meadowcroft, J. (2016). Greening Leviathan: The rise of the environmental state? *Environmental Politics, 25*(1), 1–23. https://doi.org/10.1080/09644016.2015.1085218

Dunlop, C. A., Radaelli, C. M., Wayenberg, E., & Zaki, B. L. (2024). Policy learning and policy innovation: Interactions and intersections. *Policy & Politics, 52*(4), 547–563. https://doi.org/10.1332/03055736Y2024D000000049

Eckersley, R. (2021). Greening states and societies: From transitions to great transformations. *Environmental Politics, 30*(1–2): Trajectories in Environmental Politics: 30th Anniversary Special Issue.

Fan, Y. (2010). Branding the nation: Towards a better understanding. *Place Branding and Public Diplomacy, 6*(2), 97–108.

Finnemore, M. (1996). *National interests in international society.* Cornell University Press.

Finnemore, M., & Sikkink, K. (1998). International norm dynamics and political change. *International Organization, 52*(4), 887–917.

Ganeshan, S., & Bhattacharjya, S. (Eds.). (2022). *India's role in global energy governance framework: 2040 and beyond.* The Energy and Resources Institute (TERI), in partnership with Konrad-Adenauer-Stiftung (KAS). https://www.teriin.org/sites/default/files/files/India_role_in_global_energy_GovernanceFram_work.pdf. Last accessed on 11 May 2024.

Government of India. (2008). *National Action Plan on Climate Change (NAPCC).* Prime Minister's Council on Climate Change.

Government of India. (2022, August). *India's updated first nationally determined contribution under the Paris Agreement (2021–2030).* Submission to UNFCCC. https://unfccc.int/sites/default/files/NDC/202208/India%20Updated%20First%20Nationally%20Determined%20Contrib.pdf. Last accessed on 15 May 2024.

Goyal, A. K., & Arora, S. (Eds.). (2009). *India's Fourth National Report to the Convention on Biological Diversity.* Ministry of Environment and Forests,

Government of India. https://moef.gov.in/uploads/2018/04/India_Fou rth_National_Report-FINAL_2.pdf. Last accessed on 12 August 2024.

Gupta, A., & Ghosh, S. (Eds.). (2023). *India's climate research agenda 2030 and beyond.* https://dst.gov.in/sites/default/files/India%27s%20Clim ate%20Research%20Agenda%202030%20and%20beyond.pdf. Last accessed on 14 November 2024.

Gupta, H., Kohli, R. K., & Ahluwalia, A. S. (2015). Mapping 'consistency' in india's climate change position: Dynamics and dilemmas of science diplomacy. *Ambio, 44*(4), 417–428. https://doi.org/10.1007/s13280-014-0609-5

Haas, P. M. (1992). Introduction: Epistemic communities and international policy coordination. *International Organization, 46*(1), 1–35.

Haas, P. M. (2004a). When does power listen to truth? A constructivist approach to the policy process. *Journal of European Public Policy, 11*(4), 569–592.

Haas, P. M. (2004b). Addressing the global governance deficit. *Global Environmental Politics, 4*(4), 1–15.

Haas, P. M. (2016). *Epistemic communities, constructivism and international environmental politics.* Routledge.

Hadden, J., & Seybert, L. A. (2016). What's in a norm? Mapping the norm definition process in the debate on sustainable development. *Global Governance, 22*(2), 249–268.

Harris, P. G. (Ed.). (2022). *Routledge handbook of global environmental politics.* Routledge.

Hoffmann, M. J. (2013). Global climate change. In R. Falkner (Ed.), *The handbook of global climate and environment policy* (pp. 3–18). Wiley. https://doi. org/10.1002/9781118326213.ch1

Hoppe, R. (2010). Lost in translation? A boundary work perspective on making climate change governable. In P. P. J. Driessen & P. Leroy (Eds.), *From climate change to social change: Perspectives on science-policy interactions* (pp. 109–130). International Books. https://research.utwente.nl/en/pub lications/lost-in-translation-a-boundary-work-perspective-on-making-climate. Last accessed on 10 November 2024.

Jha, V. (2023, October 18). *International solar alliance: Bridging the gap.* CESP. https://csep.org/reports/international-solar-alliance-bridging-the-gap/. Last accessed on 10 December 2024.

Katzenstein, P. J. (Ed.). (1996). *The culture of national security: Norms and identity in world politics.* Columbia University Press.

Koch, N. (2024). Green nationalism from above: Authoritarian state power and the greening of UAE nationalism. *Nations and Nationalism, 1*–20. https://doi.org/10.1111/nana.13042

Kratochwil, F. (1991). *Rules, norms, and decisions: On the conditions of practical and legal reasoning in international relations and domestic affairs.* Cambridge University Press.

Lakshmanan, P. K., Singh, S., & Lakshmi, S. (2017). Paris Agreement on climate change and India. *Journal of Climate Change, 3*(1), 1–10.

Lane, E. (2010). *How the scientific community can contribute to nuclear disarmament.* AAAS News. https://www.aaas.org/news/royal-society-aaas-how-sci entific-community-can-contribute-nuclear-disarmament. Last accessed on 15 December 2024.

Lanzerath, D., & Friele, M. (Eds.). (2016). *Concepts and values in biodiversity.* Routledge.

MEFCC. (2019). *Implementation of India's National Biodiversity Action Plan: An overview.* Ministry of Environment, Forest and Climate Change, Government of India. https://www.cbd.int/doc/world/in/in-nbsap-other-en.pdf. Last accessed on 15 October 2024.

MEFCC. (2022). *India's Long-term Low-carbon Development Strategy.* Ministry of Environment, Forest and Climate Change, Government of India. https://unfccc.int/sites/default/files/resource/India_LTLEDS.pdf. Last accessed on 10 December 2024.

MEFCC. (2023, December 12). *India at COP-28: Highlights of the 28th Conference of Parties.* Ministry of Environment, Forest and Climate Change. https://static.pib.gov.in/WriteReadData/specificdocs/documents/2023/dec/doc20231212285701.pdf. Last accessed on 10 November 2024.

Mishra, S. (2024, July). Empowering the Global South: India's role in renewable energy via International Solar Alliance (ISA). *International Journal of Research Publication and Reviews, 5*(7), 4604–4608.

Montana, J. (2021). From inclusion to epistemic belonging in international environmental expertise: Learning from the institutionalisation of scenarios and models in IPBES. *Environmental Sociology, 7*(4), 305–315. https://doi.org/10.1080/23251042.2021.1958532

Najam, A., Papa, M., & Taiyab, N. (2006). *Global environmental governance: A reform agenda.* International Institute for Sustainable Development (IISD).

Newell, P. (2000). *Climate for change: Non-state actors and the global politics of the greenhouse.* Cambridge University Press.

Nitza-Makowska, A., Longhurst, K., & Skiert-Andrzejuk, K. (2024). Green soft power? Checking in on China as a responsible stakeholder. *Polish Political Science Yearbook, 53*(1), 17–30. https://doi.org/10.15804/ppsy202402

Nye, J. S. (2004). *Soft power: The means to success in world politics.* Public Affairs.

Onuf, N. (1989). *World of our making: Rules and rule in social theory and international relations.* University of South Carolina Press.

Page, E. (2013). Climate change justice. In R. Falkner (Ed.), *The handbook of global climate and environment policy* (pp. 231–247). Wiley.

Parikh, K. S., & Parikh, J. K. (2016). Paris Agreement: Differentiation without historical responsibility? *Economic and Political Weekly, 51*(15).

Park, S. (2013). Transnational environmental activism. In R. Falkner (Ed.), *The handbook of global climate and environment policy* (pp. 268–285). Wiley. https://doi.org/10.1002/9781118326213.ch16

Park, S. (2023). The role of transnational advocacy networks in reconstituting international organization identities. *Journal of Diplomacy and International Relations, 5*(2), Article 8. https://scholarship.shu.edu/diplo_ir/vol5/iss2/8

Prasad, H. A. C., & Kochher, J. S. (2009). *Climate change and India: Some major issues and policy implications* (Working Paper No. 2/2009-DEA). Department of Economic Affairs, Ministry of Finance, Government of India. https://dea.gov.in/sites/default/files/Working%20paper%20Climate%20Change.pdf. Last accessed on 18 November 2024.

Salah, H., & Mahrous, A. A. (2019). Nation branding: The strategic imperative for sustainable market competitiveness. *Journal of Humanities and Applied Social Science, 1*(2). https://doi.org/10.1108/JHASS-08-2019-0025

Sengupta, S. (2019). India's engagement in global climate negotiations from Rio to Paris. In N. K. Dubash (Ed.), *India in a warming world: Integrating climate change and development* (pp. 142–156). Oxford University Press.

Tandon, U., Parasaran, M., & Luthra, S. (Eds.). (2017). *Biodiversity: Law, policy and governance*. Routledge.

UNEP. (2018). *Towards a green economy: Pathways to sustainable development and poverty eradication*. United Nations Environment Programme, Nairobi. https://www.unep.org/resources/report/towards-green-economy-pathways-sustainable-development-and-poverty-eradication-10. Last accessed on 15 October 2024.

Vihma, A. (2010). India and the global climate governance: Between principles and pragmatism. *Journal of Environment & Development*, 1–26.

Wendt, A. (1992). Anarchy is what states make of it: The social construction of power politics. *International Organization, 46*(2), 391–425.

Wendt, A. (1999). *Social theory of international politics*. Cambridge University Press.

Wesselink, A., Buchanan, K. S., Georgiadou, Y., & Turnhout, E. (2013). Technical knowledge, discursive spaces and politics at the science-policy interface. *Environmental Science & Policy, 30*, 1–9. https://doi.org/10.1016/j.envsci.2012.12.008

World Bank Group. (2016). *India signs deal to boost solar globally*. World Bank. https://www.worldbank.org/en/news/press-release/2016/06/30/world-bank-india-sign-deal-to-boost-solar-globally. Last accessed on 15 December 2024.

Youdon, C., & Bajaj, P. (2022, November 19). *India's approach and position on climate change governance*. https://maritimeindia.org/indias-approach-and-position-on-climate-change-governance/. Last accessed on 18 October 2024.

Zwolski, K., & Kaunert, C. (2011). The EU and climate security: A case of successful norm entrepreneurship? *European Security, 20*(1), 21–43.

CHAPTER 5

Understanding Environment and IR Theory from Non-western International Relations Theory: State-Society Interface

Vikas Kumar and Ram Babu

Abstract India is one the most important country which has set an example in wildlife conservation efforts in the world. Project Tiger, one of the most ambitious projects to save the Tiger was started in 1973, delivered fruitful results. It was not only the State which played a crucial role in this project, but also community-based traditional knowledge. In precolonial era, the local communities used to conserve the forests and wildlife with their age-old knowledge and set of traditional practices. These practices ensured the mutual coexistence of the animals and

---

ICSSR & IIT Madras, International Conference on 'Greening' of the State & Society through Securitization in India & China. Full Paper (Sub-theme: Environment and IR Theory).

V. Kumar (✉) · R. Babu
Department of Political Science, Guru Ghasidas Central University, Chhattisgarh, India
e-mail: vikaskumarpd92@gmail.com

human beings. India experienced colonial exploitation by the British. In this whole process, western knowledge systems played a very decisive role. When we talk about International Relations Theory (IRT) and episodes like this, we find a grey area which is out of the conceptual imagination what popularly known as Traditional IRT. Now, the time has come to question them.

Keywords International Relations Theory (IRT) · Tiger · Imperialism · International Relations (IR) · State · Global History · Non-Western International Relations Theory

Non-western IRT and Indian Experience of State & Society

When we talk about IRT, we always look at the problems from what is conventionally known as Traditional IRT. The Traditional IRT is plagued with war and peace, security and threat and so forth which are west-centric concepts. One of the most important drawbacks with the Traditional IRT, as Amitav Acharya would say, is State-centrism. And when we talk about the global history and the non-western experiences, the Traditional IRT falls short in grasping these experiences. What present paper does is exactly this. By putting the non-western experience within the global context, this paper tries to show that it is not the state, primarily which is shaping the environmental conservation efforts in India, but it is also the local community at large which is also playing a pivotal role. In order to understand episodes like this, we need to go beyond the dichotomy of the State and Society and look at the interface of state and society.

Sudipta Kaviraj, an Indian intellectual historian in his 'A State of Contradiction: the Post-colonial Sate in India', says that if we want to understand modernity discourse in India, it becomes necessary that we look at the relationship between modernity, which is intricately linked to the story of colonialism and along with that the instrumentality of the state in bringing that modernity to the colonial society (Kaviraj, 2010).

Kaviraj in his 'An Outline of a Revisionist Theory of Modernity' talks about some 'initial conditions' which simply happen to exist, already, as a surrounding circumstance when a historical transformation starts.

According to him, historical consciousness is always 'effective historical' which means that a particular interpretation of a text or cultural objects remains active through its effects, i.e. the effect of a particular historical reading is not really erased when it is replaced by a succeeding interpretation; effectuality of the earlier reading is never really effaced. It determines and shapes the character of the second reading and in a sense continue to exist precisely through its 'absence' (Kaviraj, 2005).

In his 'On the Enchantment of the State in India: Indian Thought on the Role of the State in the Narrative of modernity' talking about the ancient Hindu state he said God has first created Danda. Manu distinguishes between the Law (Danda) and the fallible human agent the King. He argues that the fundamental distinction between the King and the Danda leads to a theory of 'restrained rulership' and a conception of fairness of treatment towards different types of subjects. He writes that Indian society is like 'circle of circles'; each circle formed a community of neighbourhood mixed of caste, religious demonstration and occupation. While recognising the requirements of unrestricted royal authority, sought to impose restrictions upon it by positing an order that was morally transcendent- an order to which it was both subject and in complex ways eventually responsible. On the relationship between society and state he says that state will occupy the ceremonial appearance but can't interfere in the affairs of the society, which is radically different from the western conception of neutral arbiter or laissez faire (Kaviraj, 2010).

Political Theorist Bhikhu Parekh also put forth somewhat similar understanding as put by Kaviraj. In his 'Some Reflections on Hindu Political Thought', he talked about what kind of state we had in ancient Hindu society, according to him 'the king's main function was to maintain the established social order' not to change or alter it. The Hindu thinkers invented two things, the conception of Danda (discipline, force, restraint, constraint or punishment) and Dharma (holds society together). The Hindu political thinkers didn't invest the ruler with arbitrary and despotic power as the western thinkers have maintained. Here in India, the Hindu thinkers have viewed society as an organic structure of social groups, in other words 'community of communities'. Each social group, in the form of caste, had their own code of conduct and they were autonomous too. The most striking feature of the Hindu state is the status of the King. Writing about the status of the king Parekh argues 'he (the King) didn't stand above the social order. He was one of its part, but still only a part. His authority was hedged in by the relatively inviolable authority by the various autonomous centres of power, and regulated by his own specific Dharma. His authority was considerably limited by the autonomous institutions which were not his creations and had independent sources of legitimacy' (Parekh, 1986).

Migdal, Kohli and Shue in one of their books differentiated between the integrated domination and dispersed domination to explain how the situation is so complex in developing world such as India. He says in 'various settings are born the recursive relationship between state and society, the mutually transforming interactions between components of the state and other social forces. He writes that 'states are part of societies'. Once state's importance has emphasised, therefore, intellectual attention immediately shifts to issues why states differ in their respective role and effectiveness. These issues in turn can't be discussed satisfactorily without looking at society, at the socio-economic determinants of politics. So, although the important point that 'states matter' has now been made—and to repeat, it needed to be made—*there is no getting around the mutuality of the state-society interactions: societies affect states as much as, or possibly more than states affect societies'.* He farther talks about the four interrelated claims that will help in to prepare state-in-society frame of reference—first is states vary in their effectiveness based on their ties to society; second, states must be disaggregated; third is social forces, like states, are contingent on specific empirical conditions; and fourth is states and other forces are mutually empowering (Migdal et al., 1994).

In his book 'Strong Societies and weak State', Migdal defines state on the basis of its capabilities to achieve the kinds of change in society that their leaders have sought through state planning, policies and actions, according to him 'capabilities include capabilities to penetrate society, regulate social relationship, extract resources, an appropriate or use resources in determine ways'. Strong states are those with high capabilities to complete these tasks, while weak states are on the low end of a spectrum of capabilities. He also agrees on the fact that without comprehending the social structure of developing society we can't understand properly states capability. Obviously, Indian state does fall in this category of the weak state since it is not able to 'regulate some of the ill aspects of the Indian social' (for example, caste). Migdal talks about three indicators of which reflects on the social control claimed by state, namely Compliance, Participation, Legitimation. Migdal agrees that the developing countries have a web-like society and it is very hard to penetrate that, and these social structures have survived for so long because of social control dispersed among various social organisation having their own rules than centralised in the state or organisation authorised by state (Migdal, 1988).

The broader understanding of the State-Society Interface, an approach which takes cue largely from Migdal and Kaviraj. Who broadly agree on this understanding that states are best analysed as parts of societies. States may have mould, but also exchangeable moulded by, the societies within which they are embedded. States' capacity further will vary depending on their relations and ties to other social formations. And the social formations will be mobilisable in political contestations only under certain provided conditions. The political contestations pitting state against other social formations may sometimes be mutually destructive, but at other times, mutually constructive. Therefore, when we talk about IRT, it seems, there are certain phenomena which can't be grasped if we take 'state-centrism' as our reference point. In order to understand them better 'State-Society interface' would be more appropriate. The trajectories of the development of the idea of state in western world is a way different from the way it has grown in the non-western world.

GLOBAL HISTORY & IRT

Barry Buzan and Lawson in the 'Introduction' in one of their edited work talked about six assumptions regarding and the issue under consideration, first is about aspects of modernity which take maturity level in the nineteenth century; second the three interlinked process of industrialisation, the rational state and the ideologies of progress; third is their emphasise on the role played by inter-social interactions in generating the global transformation and the rejection of the view that modernity was uniquely a European phenomenon, they emphasise on the global modernity; fourth is modernity should be seen as protected uneven processes rather than a singular moment of a sharp discontinuity—there is no hard and fast distinction to be made between modern and pre-modern era; fifth, they argue that global transformation can be characterised by both the intensification of differential development and heightened interactions between societies and finally they avoid using terms like 'core' and 'periphery' the way scholars of world systems used it. According to the editors, these six assumptions produce two main claims, first is there are a set of dynamics established during the nineteenth century interlinked in a powerful configuration that reshaped the basis of international order in such a way as to define a new era, and second is this order not only transformed international relations during the long nineteenth century, it also underpins core aspects of contemporary international relation. Their contention is that the global transformation is central to understanding both the emergence of modern IR and the principal features of contemporary international order (Buzan & Lawson, 2015).

While talking about the global IR, Amitav Acharya in his 'Global International Relations and Regional Worlds: A New Agenda of International Studies' talks about idea of Global IR which aspires for greater inclusiveness and diversity within the discipline. He gives six dimensions of Global IR; first, founded on pluralistic universalism; second, founded in World History not just western history; third, it subsumes rather than supplants the existing IR theories and methods; fourth, it integrates the study of regions, regionalisms and area studies; fifth, it eschews exceptionalism; and sixth, it recognises multiple forms of agency beyond material power including resistance, normative action and local construction of global order (Acharya, 2014).

Amitav Acharya in one of his articles 'Ethnocentrism and Emancipatory IR' lists three main forms of the interrelated features of ethnocentrism which deeply work within the IRT. First and which is a common place tendency, according to him, is to simply ignore the non-western other. Second manifestation of the ethnocentrism in the IR is the tendency to view world politics from the prism of one's own national block experience and perspective or in other words a tendency to assess the other culture in terms of one's own culture. Third a form of ethnocentrism is the tendency to view the non-western experience as inferior (Acharya, 2019).

In their chapter, titled as 'World History and Development of non-Western International Relations Theory', Barry Buzan and Richard Little argued that important and necessary way for the IR theorists to make progress is to work from world history perspective. The purpose of this suggestion was that since the mainstream IRT is underdeveloped and can't reflect on certain issues especially when its conceptions of the international order brought under scrutiny. In explaining this underdevelopment of the mainstream IRT, they talk about five fundamental and interdependent shortcomings which deeply limit the potential of understanding and explaining IR, first is 'presentism', second is 'ahistoricism', third is 'eurocentrism', fourth is 'anarchophilia', and fifth is 'state-centrism' (Acharya & Buzan, 2010).

Amitav Acharya and Barry Buzan in the introduction of one of their books talked about the importance of history in the evolution of the discipline of IR. They argue that 'our overall story is this that the development of IR is actually tracks quiet closely the nature and practice of IR. Given that IR has always had strong connections to current events and foreign policy making, this link is not in itself particularly surprising. Its utility for our purposes is that it enables us to develop a nuanced insight into when, how and why acquired its notoriously west centric nature'. They further argue that 'although oversimplification, it remains broadly true that contemporary mainstream IRT is not much more than an abstraction of western history interwoven with western political theory'. They claim that if we focus on some other region, the story would be obviously different, which implies that western theories only talk about the experience of the west but not the east although they want to impose that knowledge on the rest of the world (Acharya & Buzan, 2019).

Barry Buzan and Richard Little in one of their book argued that 'the Westphalian based IRT is incapable of understanding the pre-modern international systems, but also its lack of historical perspective makes it unable to answer, or in many cases even address, the most important question about the modern international system is now so deeply ingrained that the two concepts are treated as synonymous. The whole network of terminology has grown up to reinforce this usage. At the same time the background assumption is that the international actors operate within the existing (meaning interstate) international system. We have become so inured to terminology that it is disorienting for us to think of international systems except in interstate terms. If the idea of international systems is to be extended to world history with any chance to success, it is essential to break from this association'. They further insist that the focus on the sovereign state is too limiting. The established conceptualisation only permits an understanding of Westphalian system and even for that purpose it is flawed. From this perspective they insist, the problem with IRT is that it has treated the international system standing outside history, and then used history to reinforce this ahistorical assumption. The neo-realists are particularly guilty of this fault; there is little point in turning to history with a framework that is incapable of exposing evidence of change (Buzan & Little, 2000).

NON-WESTERN CONTEXT: A CASE OF INDIA

India is a continental-sized country. It has various facets of life. On this subcontinent, civilisations has flourished, which depicted the bond among human beings, but also among human beings and animals. Since ancient times, this nation has accepted various animals as their gods and deities. The Pashupati seal was discovered in the Indus Valley Civilization's Mohenjo-daro archeological site. The seal depicts three faced god with buffalo horns reclining cross legged on a throne surrounded by an elephant, tiger and other animals. *Puli Kali* as it is popularly known in the southern state of Kerala, India, is one of the well-known forms of tiger dance around the world which is part of Indian culture. Every year people in India, almost in every state, celebrate the beginning of the festive season with *Navratre* which is the celebration of *Shero-vali-Maata* (Goddess who rides on Lions). In various parts of this subcontinent, tigers are not treated as a ferocious animal as it is depicted by the Colonial/ Imperialist people. Tiger, as an animal, is very much part of this country and culture of its people. Here, the local people have cultivated a sense of bond of mutual coexistence with them. This sense of coexistence is also visible in the national campaign to save tiger. But there is a tussle between the instrumentalities of the state, backed by scientific enquiry on the one hand, and society backed by its cultural practices and myths on the other. This tussle is not that simple that we can understand it with ease, it has its complex past, popularly known as the colonial past.

This tussle may be, in some sense, helpful in understanding the contemporary international relations. In order to understand International Relations with the conservation efforts of the flora and fauna, we need to look at the colonial history and experiences of the colonised because it is here only the real seeds of IR were sown by the dominant West.

In Nagaland, North-East India, village institutions are strong, and community-based decision making is the norm. Here indigenous communities belonging to different Naga tribes own a majority of the state's forests. This habitat connects priority wildlife populations in Northeast India with the expansive forested tracts of northern Myanmar. Working with local communities to resolve some of the needs of the locally available tiger population in the form of development of community owned piggeries, seemed an optimal strategy to provide an alternative source of protein. Working with the local communities to resolve their needs in this manner has paved the way for the WCS (Wildlife Conservation Society) India and its partners to work further with the Nagas towards strengthening wildlife corridors in the region.

In the winter of 2012, two tiger cubs were rescued in Angrim Valley village, one of the last villages on the Indo-China border in Dibang Valley of Arunachal Pradesh, India. The mother of the cubs had died and the cubs came to the village in search of food. The villagers informed this to the wildlife conservation people. The government authority wanted to convert the Dibang Wildlife Sanctuary as the Dibang Tiger reserve. But the local Mishmi people were not happy with this decision. Why was it so? Mishmi consider tigers as their elder brothers and killing a tiger is seen as a grave crime, unless human lives or their property become threatened. So this so-called rescue mission created a suspicious environment among the village people (Aiyadurai, 2016).

According to a Mishmi mythology, it is said that Mishmi and tigers were born to the same mother and were siblings, the younger hunted a deer and his (tiger) brother was eating the meat raw. He complained about his brother to his mother that his elder brother was a tiger. If he could eat raw meat, then one day he may eat the boy too. The mother came up with a plan. She made the plan in such a way that the boy emerged victorious but with a fraud. Meanwhile, the tiger was dead and his body was floating in the river water. And it is farther believed that out of his rotten carcass various other big and smaller cats took birth.

So, this is the story behind the tiger and human brotherhood. This is the story because of which humans prevent themselves from killing the tigers. For Mishmi, the tiger (*Aamra*) is their elder brother (*Apiya*). It is the most revered and feared animal, in killing it is a prohibition or in other words it is viewed as a 'homicide' (ibid.).

Angeeche, a local villager stated that and, I quote

> Why tiger reserve here? We don't hunt tigers, they are our brothers! Tigers and humans are born to the same mother. We kill tigers as only as last option, when they become a human threat or when they are killed in traps accidentally. We are protecting them anyway. (ibid.)

Many Bodo traditional ecological knowledge and skills are used by Aaranyak's 'Manas Tiger Conservation Program' to reduce pressures on tigers and their habitat in the region. There are some outreach programmes, including performing the conservation theatre inspired from popular folktale of Assam; 'Sando Bawdiani Dukhu', performed by local youths, has ignited a rich and enduring discourse among Bodo villagers about the ways, benefits and importance of living in harmony with nature. Taking a cue from traditional knowledge developed through generations by communities living with wild animals in wild will be an important link to ensuring that coexistence of human and animals (Story, 2019).

If we talk about India in the early nineteenth century, the colonial state thought that forests were unproductive. They were considered to be wilderness that had to be brought under cultivation so that the land could yield agricultural products and revenue, and enhance the income of the state. They saw the expansion of the cultivation as a sign of progress. Also, the British saw large animals as wild, primitive and savage society. They believed that by killing dangerous animals the British would civilise India. A British administrator, George Yule killed 400 tigers.

Sramek demonstrates that how tigers were closely associated with the Indian rulers, and at the same time, with all that was wild and untamed about the subcontinent. Tigers as the visual signifiers of India in its pictorial representation. Thus, the curious later Victorian and Edwardian spectacle of the British royals and other dignitaries being photographed standing aside dead tiger carcasses depicted the staging of the successful conquest of Indian nature by 'virile imperialists' (Sramek, 2006).

Susie Green in her book wrote that 'western imperialism built up a reputation for tiger that was almost entirely malign'. In western eyes, as Green argues, tiger is a 'foul, fearsome, and vicious killer' that was painted as terrible, dangerous yet a despicable beast. Western Imperial powers, the British, in Indian contexts, built a malign image of the royal Bengal tiger. Tiger's tremendous skills as an ambush predator were taken as a feature that it was a dishonourable, its tendency to revenge itself when it is wounded and persecuted to the limits of its endurance as an unquenchable desire for human flesh and blood. The local people were presented as cowed, terrified and impotent against the predations of a foul feline. In this single stroke, the imperialists thus elevated their status above that of the locals, made comfortable killings far from the elephants back into a heroic duel, legitimised their sport and made the forests easier to destroy. These heroes left buffalo calves as bait and shot tigers from the safety of their tall hideouts called *machans* (Green, 2006).

Thomas Bewick in one of his works writes about the Royal Bengal Tiger, and I quote the royal Bengal tiger is:

The most rapacious and destructive of all carnivores' animals. Fierce without provocation, and cruel without necessity, its thirst for blood in insatiable: though glutted with slaughter; it continues its carnage, nor ever gives up so long as a single object remains in its sight: flocks and herds fall indiscriminate victims to its fury: it fears neither the sight nor the opposition of man, whom it frequently makes its prey, and it is even said to prefer human flesh to that of any other animal. (Bewick, 2009)

Mackenzie writes in one of his books that if we talk about the relation between the literature on the process and projection of hunting, and the tiger seemed in some ways to be in conflict for authoritative control of the Indian environment (Mackenzie, 1988).

Significantly as Jalais notes the British were interested to show the tiger as 'worthy enemy', as it offered them the 'pleasure of measuring themselves against and 'equal' who stood his ground by virtue of his strength' which was why in the idea of iconography of popular Victorian imperialist metaphorical representations of the death of a tiger utilised as the function of showcasing the power of the Imperial British over India. As Ritvi explains, 'the connection between triumphing over a dangerous animal and subduing unwilling natives was direct and obvious' association of the game of hunter with the march of empire, was literal as well as metaphorical. In other words, the history of hunting in the Indian subcontinent in the nineteenth century was a tale of malicious masculine violence and British imperial superiority' (Crane & Fletcher, 2014).

For many twenty-first-century readers, Rudyard Kipling's Sher Khan is the tiger that immediately comes to our mind in this context. Studies of the sources of Kipling's books (1894–95) suggest that some of the settings and characters of the stories were inspired by, writes Crane and Fletcher, some of the characters and the background settings of the stories were derived from Sterndale's illustrated books such as *Seonee, or Camp Life on the Satpur Range* (1877), *Natural History of Mammalia of India and Ceylon* (1884) and *Denizens of Jungles* (1886). In the *Jungle Books*, the character of the Sher Khan is presented with such an unbridled contempt that justifies its natural enemy the man-cub Mowgli. In the last, Sher Khan dies a death he deserved, a dogs death, trampled by a herd of buffaloes when he was trapped in a ravine by Mowgli, who unwraps Sher Khan's skin and displays his 'gay stripped coat' on the council rock. In displaying Sher Khan's pelt in this way, Mowgli takes on the mantle of the British hunter, the skin reflecting the displays of the hunting trophies that adorned every club and mess in British India by the end of the nine-teenth century. Moreover, in the final decade of the century, Sher Khan's death is redolent of the 'retribution' meted out to the Indians in the wake of 1857 Uprising (ibid.).

A German expert Dietrich Brandis was appointed by the British as the first Inspector General of Forests in India. He was the one who intro-duced 'scientific forestry' in India. According to him, a proper system had to be introduced to manage the forests and people. This system would need legal sanctions. He framed the whole legal structures of forest use and 'abuse'. These forest laws changed the lives of local people in many ways. Various customary practices of the local communities were prohibited and those who were caught hunting were now punished for 'poaching'.

This colonial attitude was not only true for the Indian subcontinent but it was also true for the Central Americans too. In 1896, the American writer, Richard Harding wrote on the Honduras in the Central American that there is no more interesting question of the present day that of what is to be done with the world's land which is lying unimproved, whether it shall go to the great power that is willing to turn it to account, or remain with its original owner, who fails to understand its value. The Central Americans are like gang of semi-barbarians in a beautifully furnished house, of which they can understand neither its possibilities of comfort nor its use.

CONCLUSION

The mainstream IRT is plagued with what Robert Cox calls 'Traditional IRT'. But there are other ways too, to look at the IR, and these other ways in an umbrella identity are called 'Non-Western IRT'. Most of the third world is the result of the colonial and imperial rules. These dominating powers came from Europe and changed the future of these postcolonial nations. But their historical effect is still, in Kaviraj's language 'effective historical' which means without understanding the impact of the colonial rule and the historical experiences of the third world, it is quiet impossible to give universal theories. The so-called universal theories and paradigms, it seems, somewhere fall short to reflect on the non-western experiences. And this is the reason that a new theoretical paradigm is the need of the hour so that we will be able to understand our world in a better way.

References

Acharya, A. (2014, December). Global international relations and regional worlds: A new agenda of international studies. *International Studies Quarterly, 58*(4), 647–659.

Acharya, A. (2019). *Ethnocentrism and emancipatory IT theory.* Available on www.researchgate.net/publication/338233125. Accessed on 23/04/2022.

Acharya, A., & Buzan, B. (2010). World history and the development of non-western international relations theory. In A. Acharya & B. Buzan (Eds.), *Non-western international relations theory.* Routledge, Taylor and Francis e-Library.

Acharya, A., & Buzan, B. (2019). Introduction. In A. Acharya & B. Buzan (Eds.), *The making of the global international relations: Origin and evolution of the international relations at its centenary.* Cambridge University Press.

Aiyadurai, A. (2016). 'Tigers are our Brothers': Understanding human-animal relations in the Mishmi Hills, Northeast India. *Conservation and Society, 14*(4), 305–316, by Ashoka Trust for Research in Ecology and the Environment and Wolters Kluwer India, Pvt. Ltd.

Bewick, T. (2009). The Tiger. In T. Bewick (Ed.) *A general history of quadrupeds.* University of Chicago Press.

Buzan, B., & Little, R. (2000). Introduction. In B. Buzan & R. Little (Eds.), *International systems in world history: Remaking of the study of international relations.* Oxford University Press.

Buzan, B., & Lawson, G. (2015). Introduction. In B. Buzan & G. Lawson (Eds.), *The global transformation: History, modernity and the making of international relations.* Cambridge University Press.

Crane, R., & Fletcher, L. (2014). Picturing the Indian Tiger: Imperial iconography in the nineteenth century. In *Victorian literature and culture, 42*(3), 369–386. Cambridge University Press.

Green, S. (2006). The psychology of fear: The tiger tamed, the tiger degraded. In S. Green (Ed.), *Tiger.* Reaktion Books.

Kaviraj, S. (2010). On the enchantment of the state in India: Indian thought on the role of the state in the narrative of modernity. In S. Kaviraj (Ed.), *Trajectories of Indian state: Politics and ideas.* Columbia University Press, Permanent Black.

Kaviraj, S. (2005). An outline of a revisionist theory of modernity. *European Journal of Sociology, 46*(3), 497–526.

Mackenzie, J. M. (1988). The Imperial Hunt in India. In J. M. Mackenzie (Ed.), *The empire of nature: Hunting, conservation, and British imperialism.* Manchester University Press.

Migdal, J. (1988). Introduction. In J. S. Migdal (Ed.), *Strong societies and weak states: State-society relations and the state capabilities in the third world.* Princeton University Press.

Migdal, J. S., Kohli, A., & Shue, V. (1994). Introduction: Developing Sate-in- Society perspective. In J. S. Migdal, A. Kohli, & V. Shue (Eds.), *Sate power and social forces: Domination and transformation in the third world*. Cambridge University Press.

Parekh, B. (1986). Some reflections on the Hindu tradition of political thought. In T. Pantham & K. L. Deutsch (Eds.), *Political Thought in Modern India*. Sage Publications.

Sramek, J. (2006, Summer). Face him like a Briton": Tiger hunting, imperialism, and British masculinity in colonial India. In *Victorian Studies, 48*(4), 659–680, Indiana University Press.

Story, IUCN. (2019). Community involvement: the key to successful Tiger Conservation. Online available on https://www.iucn.org/news/species/201 908/community-involvement-key-successful-tiger-conservation. Accessed on 19/02/2024.

PART III

Trends in Environmental Securitisation

Identity Politics and Chinese Climate Securitisation

Juha A. Vuori

Abstract This chapter analyses the Chinese discourse on global climate change and relates it to the PRC's identity politics. As such, the issue of climate change in security studies has evolved from debates about environmental security. The paper relates the PRC's macropoliticisation of climate change to its longer-view approach to the treatment and role of the environment in China. This shows how environmental concerns have raised high on the discursive political agenda, yet how climate change is regarded more as an issue of international politics than national security in the PRC. This evolution becomes understandable when viewed in terms of identity politics.

The chapter is a shortened and slightly modified version of a chapter in Vuori (2024).

J. A. Vuori (✉)
Department of Philosophy, Contemporary History, and Political Science, University of Turku, Turku, Finland
e-mail: juha.vuori@utu.fi

J. T. Karackattu et al. (eds.), *Environmental Securitisation in India and China*, https://doi.org/10.1007/978-981-97-9160-6_6

97

Keywords China · Climate Change · Securitisation · Politicisation · Identity Politics

INTRODUCTION

Global climate change can be considered an international macrosecuritisation discourse (Buzan & Wæver, 2009): it represents a physical threat universalism that concerns most of humanity. As such, climate change can be included as a subset of environmental politics. While some forms of environmental degradation can be catastrophic locally, they may not be of global concern. At the same time, some emissions that contribute to climate change may not be an issue locally yet may end up having global repercussions. Such features have made the securitisation of climate change a contentious issue in many places worldwide. The People's Republic of China (PRC) is currently the world's largest source of carbon emissions into the atmosphere. It has been actively engaged in international climate diplomacy but has not been a vocal securitising actor. The role of environmental protection has gained its position on the political agenda during the Xi Jinping administration. Still, the issue of climate change has its particular trajectory in the PRC.

As such, climate change has been advocated as a global, or in some places, national security issue by numerous NGOs, as well as by the Fourth Assessment Report of the IPCC in 2007. Climate change has also been on the agenda of the UNSC, where 'everyone's future' was claimed to be at stake (UNSC 2007: 35; Bothe, 2008; for development in the UNSC, see Hardt et al., 2023). States and institutions vary in how they view and value issues like climate change. For example, global climate change is a security agenda item first and a developmental concern second for the European Union's (EU's) external action service (Odeyemi, 2021). In contrast, for the PRC, climate change is an issue of development first and security second.

Indeed, the PRC was less inclined to phrase climate change in security terms in the 2000s. Indeed, it has been against dealing with the issue within the UNSC (Vuori, 2023), favouring the United Nations Framework Convention on Climate Change (UNFCCC) instead. For a long time, the PRC refused to bind itself with international emission reduction commitments. However, both positions changed in the latter half of the 2010s, and care for the environment has become a core feature of Xi Jinping's ideological thought. The present chapter connects the evolution of China's position in environmental and climate politics to its identity politics.

Identity and Environmental Politics in Mao's China

Chinese philosophy and many imperial-era rulers revered nature and even connected to the 'mandate of heaven' to its favourable or disastrous features (Harris, 1976). In contrast, Mao Zedong's approach to the environment was quite militaristic: rather than something to be protected, nature was to be conquered in mass campaigns and massive construction projects (Shapiro, 2001: 3–4). Mao can even be characterised as having waged a domestic 'war on nature' (Shapiro, 2001). The goal was to increase China's population and harness natural resources to serve national reconstruction and improve the PRC's standing worldwide.

China's international environmental politics started in 1972 with participation in the UN Conference on the Human Environment. In line with its overall identity politics, the PRC's position was to consider environmental degradation the result of imperialism and capitalism, emphasise the industrialised or developing nation dichotomy, and be adamant that international agreements could not jeopardise the PRC's sovereignty or economic development. For example, the delegate at the 1972 UN conference pointed out the main reason for pollution as 'the policy of plunder, aggression and war carried out by imperialist, colonialist and neocolonialist countries, especially by the superpowers' (UN, 1972: 63). The PRC also opposed the linking of population growth with environmental degradation and food shortages and began promoting its long-standing stance of 'common but differentiated responsibilities' between developed and developing nations regarding the environment (which was still present in the PRC's statement at the UNSC in 2019). This initial meeting is where the PRC's long-term policy line was formed: developing nations should guard their independence and not allow environmental issues to hamper their development goals (Kopra, 2016b: 158).

IDENTITY AND ENVIRONMENTAL POLITICS
IN THE REFORM ERA AND THEREAFTER

As with many other aspects of the PRC's politics, the two-pronged Maoist legacy informed its environmental and climate politics in the reform era. Indeed, economic growth largely trumped other concerns in the environmental and, later, climate field. Like in many other policy areas in post-Mao China, a gradual change in the political leadership's approach to the environment is discernible.

The issue of climate change rose to prominence in the international arena in the early 1990s, which was a time when the PRC found itself in relative diplomatic isolation after the violent events in 1989 (Vuori, 2003, 2018). This is a partial explanation for the PRC's signing and ratification of the Kyoto Protocols, something that has been described as a triumph for its foreign policy from the viewpoint of its national interests (Chen, 2008: 150): the protocol only required emission cuts from developed states, which the PRC was not included in. With the ratification of this major climate policy treaty, the PRC could provide some credibility for its campaign to present itself as a 'responsible great power' (Kopra, 2016b).

At the highest level of political line formation, Jiang Zemin (1997) was the first to mention the environment in a report at the Party Congress in 1997 and to finally recognise the connection between environmental strain and population growth that Mao's China had denied. Jiang's successor, Hu Jintao (2007), set the building of an 'ecological civilisation' as a goal at the 17th Party Congress in 2007. The notion was added to the constitution of the Communist Party at the 18th Party Congress in 2012, and environmental damage and ecological benefits were made assessment criteria in the career development of state officials (Hu, 2012). A further step was the establishment of the Environmental Police as a watchdog department under the Environmental Protection Bureaus for the enforcement of environmental protections (Joseph & Karackattu, 2022.

In succinct terms, the PRC's long-term position within the international field of environmental politics in the reform era can be described as the supremacy of sovereignty, development first and emphasis on the industrialised or developing nation dichotomy (Chen, 2012). The PRC became a net oil importer in the early 1990s, which made energy security more important. Dependence on oil imports has been a perennial cause for concern ever since, as imports exceeded domestic production in

2009 (Ghiselli, 2021: 75–76; Nyman & Zeng, 2016: 303). This combination of concerns produced the paradoxical situation where the PRC was against internationally binding emission cuts when it made major strides in the development and deployment of emission-reducing technologies and increased the supply of non-fossil fuel energy (Chen, 2012; Dent, 2014). Due to its reliance on maritime supplies for its oil imports, realpolitik and green goals began to become aligned in China.

The Paris Climate Agreement 2015 was a major turning point in China's international climate politics, as the PRC committed itself to binding cuts for the first time in a multilateral agreement (Kopra, 2019: 110). In preparing for this transition, the PRC made efforts to turn its climate policy into a benefit for the economy rather than a drag. These have aimed to decouple emissions and economic growth (Hernandez & Misalucha-Willoughby, 2020), and some even see this as the securitisation of economic development (Sahu, 2021). The PRC's lead in the production of wind and solar energy power plants is an indicator of success in this policy line.

On the domestic level, Xi Jinping continued promoting an ecological civilisation in the PRC as a major strategic goal, with green development as crucial. The green GDP was relaunched in 2015 (Gongsheng et al., 2015) as part of the policy line of a 'new era of socialist ecological civilisation.' More importantly, climate change has been incorporated into the main policy line that has become the mainstay of Xi's ideological formulation of 'socialism with Chinese characteristics' that entails many societal aspects by the National Development and Reform Commission (NDRC).

In other words, environmental concerns should be included in all aspects of the ideal of Chinese society, economy, and politics, into the main ideological concept of Xi Jinping's thought: protecting the environment and harmony between humans and nature are part of socialist civilisation in this new ideological formulation. In this vision, nature is to be guarded rather than fought against, and indicates that the Maoist tradition regarding the environment has been abandoned since the 2010s, and the PRC under Xi has continued to emphasise a 'community of a shared future of humankind' (Hu, 2012; Xi, 2017a). However, climate change is a separate concern that relates to development.

Indeed, there has been a significant change in terms of politicising nature in the PRC. In Mao's China, nature was to be conquered for economic and other development. In contrast, today, 'harmony between human and nature' (Xi, 2017b: 20–22) is an essential part of the current ideological formation that guides the Communist Party and the People's Republic. This position and commitment to green development and 'harmony between humans and nature' were reiterated in Xi's (2022) report to the Party Congress as well (Xi, 2022), even if it was not the main thrust of the report.

This shift has had effects in practice too. At the time of writing, the PRC is now the world's top producer of renewable energy. Satellite images have revealed many enormous solar and wind farms across the Gobi Desert. Chinese companies now dominate the global market for solar panels and wind turbines (UNEP, 2022). Such developments signal a strong commitment to green politics. Yet, Chinese state companies have announced controversial decisions to expand the use of coal in energy generation in 2023 (Bloomberg, 2023). This would indicate that energy self-sufficiency still trumps concerns for the shared future of humankind.

IDENTITY, DEVELOPMENT AND CLIMATE SECURITY

The PRC's discourse on climate change has gradually shifted from developmental concerns towards considering the issue in terms of security. Still, in 2007, the NDRC framed the issue in terms of 'impact' and 'development' (NDRC, 2007). In 2008, the white paper on climate change (SCIO, 2008) indicated that the PRC was more willing to align itself with the general international trend on issues of climate change by phrasing the issue in terms of threat rather than impact. Still, the proposed means to tackle it were closer to macropoliticisation than macrosecuritisation (Buzan & Wæver, 2009; Vuori, 2011). In other words, the issue was raised as one of concern but not one of survival that required drastic action. The reason for this challenge to the survival and development of society is placed on the activities of developed nations (SCIO, 2008), as was carried out in Mao's China (Kopra, 2019). The PRC was depicted as a developing nation, which is adversely affected by climate change that threatens its 'natural ecosystems as well as the economic and social development' (SCIO, 2008). Still, the suggested measures were not extraordinary or exceptional in the way security politics tends to be understood (Buzan et al., 1998) and were in accordance with the then prominent foreign policy line of 'harmonious development' (Vuori, 2015). In terms of securitisation theory, these kinds of formulations can be considered politicisation rather than securitisation.

The year 2012 was crucial for the PRC's climate politics. The PRC identified the dire effects of climate change on its domestic situation and presented itself as among the states most vulnerable to the adverse effects (NDRC, 2012, 2013). Yet, the measures to combat such effects were not legitimated with security logic. Accordingly, in international climate change forums, the PRC's line was still the promotion of the 'principles of fairness and "common but different responsibility"' and to actively safeguard the interests and legitimate development rights of developing countries (NDRC, 2012: 24). In the domain of South–South cooperation in the mitigation of climate change, the PRC claimed to operate 'based on the principle of "mutual benefit and win–win cooperation, and being practical and effective"' (NDRC, 2013: 60).

While the PRC and the US were on the opposite side of the development divide, their policies had been fairly similar in the climate negotiations: avoid constraints on their domestic economy that binding commitments to emission reductions would bring (Nyman, 2018). The similar tones among these states in their climate policies were also evident in their joint declarations on climate change (State Department of the United States of America, 2013; White House Office of the Press Secretary, 2014). Here too, the means promoted to deal with the danger of climate change fell within overall international politics, not the realm of security. Indeed, rather than an urgent issue of security, 'tackling climate change' was seen to 'strengthen national and international security' (White House Office of the Press Secretary, 2014).

The PRC's foreign policy in the 2000s has consistently aimed to avoid the impression that it would be a threat, even as its power resources increase (Vuori, 2018): 'China's development does not threaten any other country. No matter what stage of development it reaches, China will never seek hegemony or engage in expansion' (Xi, 2017b: 53). The environment has not been an exception here: the assurances that the PRC does not pose an environmental threat to the world have been part of the PRC striving to present itself as a responsible great power (Kopra, 2016a: 20). It seems that states' international images are an important facet of their climate policy: action to curtail climate change by the U.S. and China is crucial to set a 'powerful example that can inspire the world' (State Department of the United States of America, 2013). From the viewpoint of the PRC, climate change as foreign policy combines the issues of soft power (image as a responsible power in the climate field) with harder forms of power (economic growth cannot be jeopardised) (Chen, 2012: 106).

The joint Sino-U.S. efforts were an augur for the success of the 2015 Paris Climate Conference. President Trump's decision to withdraw from the Paris Agreement allowed the PRC to enforce its image as a responsible great power. While its commitments were not at the level of, for example, Europe, the PRC retaining its line with the U.S. withdrawal provided for a positive image, nevertheless. The PRC had previously changed its policy in tandem with the U.S., which had empowered the PRC, raised its standing, promoted the status of a responsible great power and had not weakened its economic standing in relation to U.S. industries. The embodiment of the climate line here was part of Xi's overall policy line of the PRC as a responsible great power.

Indeed, as Trump initiated the U.S. withdrawal process from the Paris Agreement, Xi Jinping described the PRC's role as a 'torchbearer' in the global response to climate change in his speech at the 19th Party Congress (Xi, 2017b). In the 2019 report on China's climate change policies, the PRC supported 'the comprehensive and effective implementation of the Paris Agreement' (MEE, 2019: 29). Yet, while the PRC departed from the U.S. position in this regard, it did not quite manage to fill the vacuum left by the U.S. by making new initiatives in international negotiations (Kopra, 2019: 148; Zhang & Orbie, 2019: 20). Furthermore, the Biden administration returned the U.S. to the Paris Agreement and appeared to be taking a more climate-friendly stance overall compared with the Trump administration. This may be why the U.S. and the PRC made a new joint statement during the climate summit 2021 (Department of State of the United States of America, 2021).

The PRC's transition towards binding international commitments was first evident in the domestic discussion, where climate change was presented as a direct threat and as having implications through its indirect effects in other security-related fields, such as social stability, which was already under stress from the major health issue of air pollution. This was a major impetus for Premier Li Keqiang's declaration of a 'war on pollution' in 2014 (Reuters, 2014). As such, ecological security was included in the national security system for the first time in the first meeting of the Central National Security Commission (CNSC) (*Xinhua*, 2014a). In its entirety, Xi's 'holistic,' integrated, 'overall security outlook' or 'national security path with Chinese characteristics' listed 11 issue areas of concern: 'the spheres of politics, territory, military, economy, culture, society, science and technology, information, ecology, nuclear, and natural resources' (*Xinhua*, 2014a).

Xi solidified the line of the 'beautiful China initiative' and emphasised 'global ecological security' in the 19th Party Congress in 2017 (Xi, 2017b). Indeed, noting the progress made in building an ecological civilisation was among the first categories he reported on in the speech. Beyond its domestic efforts, Xi noted that the PRC has taken the 'driving seat in international cooperation to respond to climate change' and 'become an important participant, contributor, and torchbearer in the global endeavour for ecological civilization' (Xi, 2017b: 4). Furthermore, 'ensuring harmony between human and nature' is a part of 'socialism with Chinese characteristics in the new era,' as is a 'holistic approach to national security' that includes elements, such as 'traditional and non-traditional security, and China's own and common security' that aim to 'foster new thinking on common, comprehensive, cooperative, and sustainable security' (Xi, 2017b: 20–22). Climate change is also listed among the common uncertainties and destabilising factors that humanity faces (Xi, 2017b: 52): 'unconventional security threats like terrorism, cyber-insecurity, major infectious diseases, and climate change continue to spread.' To 'build a community with a shared future for mankind,' 'we should be good friends to the environment, cooperate to tackle climate change, and protect our planet for human survival' (Xi, 2017b: 53).

While ecology and conservation of nature and the fight against various forms of pollution were mentioned in several sections of the speech, national security and military issues had their sections. Indeed, the presentation of the most vital national security issues does not explicitly contain the climate (Xi, 2017b: 20–21).

Xi (2022) continued to emphasise the holistic concept of national security in his 2022 report to the Party Congress. The climate was not part of the explicit elements of national security here either: 'We must take the people's security as our ultimate goal, political security as our fundamental task, economic security as our foundation, military, technological, cultural, and social security as important pillars, and international security as a support.' While the climate is included within 'non-traditional security,' it is covered in the report in its non-security section. Accordingly, climate change was presented as a concern for shared human survival, not so much as a direct threat to the national security of the PRC (Xi, 2022).

This indicates that climate change is recognised as a security issue. Still, the referent object is humanity more than the Chinese nation, and, therefore, the ways to tackle it fall on international efforts where the PRC is also increasing its leadership role. The PRC's national security focusses on political security, separatism and terrorism (Vuori, 2024). At the same time, environmental civilisation has become an integral part of the overall ideology of socialism with Chinese characteristics. While this may result from security-oriented thinking, it is not legitimised with national security speech.

CONCLUSIONS: FROM A WAR ON NATURE TO HARMONY BETWEEN HUMANS AND NATURE

The PRC's reluctance to bind itself to international commitments in the climate field is understandable, as it coheres with the country's overall approach to international politics over the last 40 years. The PRC's leaders have emphasised the PRC's need to concentrate on its internal development and the creation of a peaceful zone around its borders. This has enabled China's rise and return to being a major power in world politics and has been called 'the period of strategic opportunity.' As the PRC has become more affluent, and the end of the Cold War ended block politics in the PRC (Vuori, 2024), multilateral diplomacy has become the norm in its international politics. The aim here has been to ensure that international organisations are not used against the PRC's interests and that it can affect the creation of international norms, which is termed 'discourse power' by its leadership (e.g. *The People's Daily*, 2016). As the PRC is dependent on foreign oil imports, it has been more concerned with energy security and sustained development than with the growing issue of climate change. As its overall foreign policy line has shifted towards recognising the PRC as a major power with 'responsibilities' (Kopra, 2016b, 2019), the international approach to climate agreements has also changed.

Sovereignty, development and energy concerns have trumped climate security in the PRC. Since the mid-2010s, the PRC has committed itself to internationally binding climate actions if they are not externally imposed. This has happened in tandem with a change in the PRC's overall foreign policy stance, where it is portraying itself as more of a responsible great power than a developing nation. It still maintains the principle of equity and common but differentiated responsibilities respective to capabilities when responding to climate change, as it did in the most recent UNSC debates in 2020 and 2021 (Vuori, 2023). Still, the PRC has performed more climate actions domestically than it has committed to internationally, and it made domestically binding decisions on emission cuts before doing them internationally. Environmental damage and recovery are criteria in the evaluation of state officials. Unfortunately, the central policies may be more progressive than what is implemented locally due to the prevalent issue of corruption.

In the PRC's current policy documents, climate change is listed among a number of issues that challenge human survival on a global scale. Environmental and climate security is part and parcel of the current overall security outlook that emphasises collective and universal security that includes 'human security' with Chinese characteristics (Breslin, 2015). Regarding national security, political security, terrorism and separatism are the main concerns (Vuori, 2024). Climate change affects human security in terms of social development, lives and property. Yet, 'harmony between humans and nature' goes beyond security because it is a vital aspect of 'socialism with Chinese characteristics in the new era,' which is an essential part of Xi Jinping's thought. The climate issue is also beneath the umbrella conceptualisation of the community of a shared future for humankind (Vuori, 2024).

From the viewpoint of macrosecuritisation (Buzan & Wæver, 2009), the referent objects for the danger of climate change include national and global levels (Table 6.1). In addition, the danger is an encompassing one, where some of the PRC's most crucial national interests may be endangered. In terms of political moves, the PRC seems to be promoting the macropoliticisation of global climate change and has infused the climate issue within the core of its political agenda, albeit without legitimising this with national security. The most recent development suggests that the PRC is willing to recognise that the issue can be a national security concern in some parts of the world, even within the UNSC (Vuori, 2023). As such, the PRC's discourse presents referent objects at national and global levels in terms of physical threat universalisms in various forms. At the same time, the danger is framed as an encompassing one, where some of the PRC's most crucial national interests may be endangered. As seen elsewhere (Buzan & Wæver, 2009), physical threat universalisms as macro issues appear less effective in political mobilisation than the universalisms that produce a matched pair.

Table 6.1 Macrosecuritisation Elements of China's climate change discourse

Macrosecuritisation discourse	Types of moves	Type of universalism	Alignment in constellations	View on polarity	Bureaucratic logic
Global climate change	Politicisation	Physical threat (and economic problem)	Developing states versus industrialised nations; responsible great power	Multipolar	Environment, diplomacy and economy

The issue of climate change is part of the PRC's contemporary identity politics in two ways. First, the way the PRC has emphasised the common but differentiated responsibilities of industrialised and developing nations in handling the issue has maintained the line the PRC adopted in international environmental politics in the 1970s: environmental protection, and later the mitigation of climate change, cannot encroach on either the development of the PRC or its sovereignty. The early stage of international climate politics also allowed the PRC to maintain its identity as a developing nation, even though many indicators indicated something completely different. Second, climate politics have allowed the PRC to promote its new identity as a 'torchbearer' or a new kind of 'responsible great power' that is great but not an irresponsible and aggressive imperialist like the great powers during the Cold War.

Beyond issues of identity, the PRC has developed a resilience strategy when combating natural disasters (Ministry of Emergency Management, 2022). This suggests that the response to the challenge of climate change could take the form of mitigation rather than prevention in the PRC. Still, resilience can evolve into securitisation (Bourbeau & Vuori, 2015). Accordingly, it is possible that the PRC's resilience strategy could lead to the securitisation of climate change on a national level, especially when scientists link the flooding of rich regions along the Yangtze with melting glaciers on the plateau.

The measures that the PRC has proposed internationally are not extraordinary nor exceptionalist. This is so even after it conceded that climate change could undermine peace and stability and that it concerns the UNSC from the viewpoint of peace and security (Vuori, 2023). Rather, the measures the PRC has proposed emphasise cooperation, multilateralism, development and peace. While the PRC's current emission reduction goals are not overly ambitious, environmental issues, including climate change, have become more and more prominent in its overall policy doctrines and ideology since Hu Jintao's administration and are even more deeply integrated into Xi Jinping's ideological guidelines. As Xi Jinping has removed term limits from his leadership positions, this trajectory will most likely continue with him as leader.

References

Bloomberg. (2023). *China to speed up construction of coal power plants this year*. https://www.bloomberg.com/news/articles/2023-01-20/china-to-speed-up-construction-ofcoal-power-plants-this-year Accessed 21 Feb 2023.

Bothe, M. (2008). Security in international law since 1990. In H. G. Brauch et al. (Eds.), *Globalization and environmental challenges. Reconceptualizing security in the 21st century* (pp. 475–485). Springer-Verlag.

Bourbeau, P., & Vuori, J. A. (2015). Security, resilience and desecuritization: Multidirectional moves and dynamics. *Critical Studies on Security*. https://doi.org/10.1080/21624887.2015.1111095

Breslin, S. (2015). Debating human security in China: Towards discursive power?

Buzan, B., & Wæver, O. (2009). Macrosecuritisation and security constellations: Reconsidering scale in securitisation theory. *Review of International Studies*. https://doi.org/10.1017/S0260210509008511

Buzan, B., Wæver, O., & de Wilde, J. (1998). *Security*. Lynne Rienner Publishers.

Chen, G. (2008). China's Diplomacy on climate change. *The Journal of East Asian Affairs, 22*(1), 145–174.

Chen, G. (2012). *China's climate policy*. Routledge.

Dent, C. M. (2014): *Renewable energy in East Asia: Towards a new developmentalism*. Routledge.

Department of State of the United States of America. (2021). U.S.–China joint Glasgow declaration on enhancing climate action in the 2020s. https://www.state.gov/u-s-chinajoint-glasgow-declaration-on-enhancing-climate-action-in-the-2020s/ Accessed 21 Feb 2023.

Ghiselli, A. (2021). *Protecting China's Interests overseas: Securitization and foreign policy*. Oxford University Press. *Journal of Contemporary Asia*. https://doi.org/10.1080/00472336.2014.907926

Gongsheng, P., Jun, M., & Zadek, S. (2015). Establishing China's green financial system: Final report of the green finance task force. *People's Bank of China, United Nations Environment Programme.* https://wedocs.unep. org/bitstream/handle/20.500.11822/8459/-Establishing_China%E2%80% 99sGreen_FinancialSystem-2015PBC_UNEP_InquiryGreen_Task_Force_R epor.pdf?sequence=4&%3BisAllowed=y%2C%20Chinese%7C%7CHttps%3A// wedocs.unep.org/bitstream/handle/20.500.11822/8459/Estab

Hardt, J. N., Harrington, C., Simpson, N., von Lucke, F., & Estève, A. (2023). *Climate security in the anthropocene: Exploring the approaches of United Nations security council member-states.* Springer.

Harris, N. (1976). *Mandate of heaven.* Quartet Books.

Hernandez, A. M., & C. Misalucha-Willoughby. (2020). Securitization of climate and environmental protection in China's new normal. *Decision-Making in Public Policy & the Social Good eJournal.* https://doi.org/10.2139/ssrn.368 6466

Hu, J. (2007). Hold high the great banner of socialism with Chinese characteristics and strive for new victories in building a moderately prosperous society in all respects. *China Daily.* https://www.chinadaily.com.cn/china/ 19thcpcnationalcongress/2010-09/07/content_29578561.htm. Accessed 15 Feb 2023.

Hu, J. (2012). *Firmly March on the path of socialism with Chinese characteristics and strive to complete the building of a moderately prosperous society in all respects.* https://www.china.org.cn/china/18th_cpc_congress/2012-11/16/ content_27137540.htm. Accessed 15 Feb 2023.

Information Office of the State Council of the People's Republic of China (SCIO). (2008). *China's Policies and Actions for Addressing Climate Change.* https://www.china.org.cn/government/whitepaper/node_ 7055612.htm Accessed 29 Oct 2008.

Jiang, Z. (1997). Hold high the great banner of Deng Xiaoping theory for an all- round advancement of the cause of building socialism with Chinese characteristics' into the 21st century. Report Delivered at the 15th National Congress of the Communist Party of China on September 12, 1997. *Beijing Review.* https://www.bjreview.com.cn/document/txt/2011-03/25/ content363499.htm. Accessed 16 Feb 2023.

Joseph, J., & Karackattu, J. T. (2022). State actions and the environment: Examining the concept of ecological security in China. *Environment, Development and Sustainability, 24*(11), 13057–13082.

Kopra, S. (2016b). *With great power comes great responsibility? China and the international practice of climate responsibility.* Doctoral dissertation. Tampere University. https://trepo.tuni.fi/handle/10024/99873

Kopra, S. (2016a). Great power management and China's responsibility in international climate politics. *Journal of China and International Relations.* https://doi.org/10.5278/OJS.JCIR.V4I1.1513

Kopra, S. (2019). *China and great power responsibility for climate change*. Routledge.

Ministry of Ecology and Environment (MEE) of the People's Republic of China. (2019). *China's policies and actions for addressing climate change*. https://www.ncsc.org.cn/yjcg/cbw/202111/P020211117418821628432.pdf

Ministry of Emergency Management. (2022). *National strategy for prevention and reduction of natural disasters*. National Committee for Reduction of Disasters.

National Development and Reform Commission of the People's Republic of China (NDRC). (2007). *China's national climate change programme*. http://english.mee.gov.cn/Resources/Plans/Plans/200710/P020071016292571780686.pdf. Accessed 17 Feb 2023.

National Development and Reform Commission of the People's Republic of China (NDRC). (2012). *China's policies and actions for addressing climate change*. https://www.ncsc.org.cn/yjcg/cbw/201307/W020180920484675924333.pdf. Accessed 17 Feb 2023.

National Development and Reform Commission of the People's Republic of China (NDRC). (2013). *China's policies and actions for addressing climate change*. http://img.thupdi.com/news/2015/08/1440491483531951.pdf. Accessed 17 Feb 2023.

Nyman, J., & Zeng, J. (2016). Securitization in Chinese climate and energy politics.

Nyman, J. (2018). *The energy security paradox. Rethinking energy (in)security in the United States and China*. Oxford University Press.

Odeyemi, C. I. (2021). *Climate risk and climate security: A Comparison of norm emergence under the FCCC, the EU and the UNSC, 2001–2019*. Doctoral dissertation. University of Technology, Sydney. http://hdl.handle.net/10453/149345

People's Daily. (2016). 调动各方面力量参与全球经济治理 协力提高制度性话语权(新知新觉) [Improve China's institutional discourse power in global economy governance]. http://opinion.people.com.cn/n1/2016/0219/c1003-28134857.html. Accessed 28 Mar 2021.

Reuters. (2014). *China to 'Declare War' on pollution, premier says*. https://www.reuters.com/article/us-china-parliament-pollution-idUSBREA2405W20140305. Accessed 17 Feb 2023.

Sahu, A. K. (2021). From the climate change threat to the securitisation of development: An analysis of China. *China Report*. https://doi.org/10.1177/00094455211004259

Shapiro, J. (2001). *Mao's war against nature: Politics and the environment in revolutionary China*. Cambridge University Press.

State Department of the United States of America. (2013). Joint U.S.–China statement on climate change. https://2009-2017.state.gov/r/pa/prs/ps/2013/04/207465.htm. Accessed 16 Feb 2023.

United Nations. (1972). *Report of the United Nations Conference on the Human Environment, Stockholm, 5–16 June 1972.* A_CONF.48_14_Rev.1-EN. https://digitallibrary.un.org/record/523249/files/A_CONF.48_14_Rev.1-EN.pdf?ln=en

United Nations Security Council. (2007). *Security council holds first- ever debate on impact of climate change on peace, security, hearing over 50 speakers.* SC/9000, 17.4.2007. https://press.un.org/en/2007/sc9000.doc.htm. Accessed 20 Feb 2023.

United Nations Environment Programme (UNEP). (2022). *Renewables 2022 global status report.* UNEP.

Vuori, J. A. (2011). *How to do security with words? A grammar of securitisation in the People's Republic of China.* Doctoral dissertation. University of Turku. https://www.utupub.fi/handle/10024/70743

Vuori, J. A. (2015). Climate politics in Chinese foreign policy. In J. Y. Cheng & M. Siika (Eds.), *New trend and challenges in China's foreign policy* (pp. 227–250). City University of Hong Kong Press.

Vuori, J. A. (2023). Climate security in China: An issue for humanity rather than the Nation. In J. N. Hardt & C. Harrington, F. von Lucke, A. Estève & N. Simpson (Eds.), *Climate security in the anthropocene: Exploring the approaches of United Nations security council member-states* (pp. 45–63). Springer.

Vuori, J. A. (2003). Security as justification: An analysis of Deng Xiaoping's speech to the Martial Law Troops in Beijing on the ninth of June 1989. *Politologiske Studier, 6*(2), 105–118.

Vuori, J. A. (2018). Let's just say we'd like to avoid any great power entanglements: Desecuritization in Post- Mao Chinese foreign policy towards major powers. *Global Discourse.* https://doi.org/10.1080/23269995.2017.1408279

Vuori, J. A. (2024). *Chinese macrosecuritization: China's Alignment in global security discourses.* Routledge.

White House Office of the Press Secretary. (2014). *U.S.–China joint announcement on climate change.* https://obamawhitehouse.archives.gov/the-press-office/2014/11/11/us-china-joint-announcement-climate-change. Accessed 20 Feb 2023.

Xi, J. (2017b). Secure a Decisive victory in building a moderately prosperous society in all respects and strive for the great success of socialism with Chinese characteristics for a new era.

Xi, J. (2017a). Work together to build a community of shared future for mankind.

Xi, J. (2022a). *Hold high the great banner of socialism with Chinese characteristics and strive in unity to build a modern socialist country in all respects.* Report to the 20th National Congress of the Communist Party of China. https://www.mfa.gov.cn/web/system/index_17321.shtml. Accessed 20 Feb 2023.

Xinhua. (2014a). 习近平就云南昆明火车站暴力恐怖案件作出重要指示- 高层动态-新华网 [Xi Jinping Issues important directives following the violent terrorist incident at the Kunming railway station in Yunnan]. https://www.xinhuanet.com//politics/2014-03/02/c_119564039.htm Accessed 27 Feb 2023.

Xinhua. https://www.xinhuanet.com/english/2017-01/19/c_135994%20707.htm Accessed 20 Feb 2023.

Zhang, Y., & Orbie, J. (2019). Strategic narratives in China's climate policy: Analysing three phases in China's discourse coalition. *The Pacific Review.* https://doi.org/10.1080/09512748.2019.1637366

Environmental Policy in China and Interaction with Border Countries (the Case of Russia)

Olga Zalesskaia ⓘ

Abstract Today, China actively participates in numerous international environmental forums and conferences. Environmental cooperation is an important part of Russian-Chinese cooperation. Russia and China pay attention to environmental policy issues, but these issues are not a priority, much more attention is paid to cooperation in the economic, energy, political, financial and military spheres. The study delves into the collaborative efforts between Russia and China in the transboundary regions of the Far East to address environmental concerns. The study examines the challenges and opportunities for environmental cross-border cooperation in the twenty-first century. Specifically, it focusses on the joint initiatives taken by Russia and China in the border Far Eastern territories to prevent environmental incidents and implement remedial measures. By emphasising the significance of bilateral efforts in safeguarding the environment

O. Zalesskaia (✉)
Faculty of Foreign Languages, Blagoveschensk State Pedagogical University, Blagoveshchensk, Russia
e-mail: olgazalesskaya@gmail.com

© The Author(s), under exclusive license to Springer Nature Singapore Pte Ltd. 2024
J. T. Karackattu et al. (eds.), *Environmental Securitisation in India and China*, https://doi.org/10.1007/978-981-97-9160-6_7

115

for future generations, the article aims to inspire further dialogue and action towards sustainable practices in the Far East region.

Keywords China · Russia · The border Far Eastern Territories · Environmental Cooperation · Environmental Threats · Regional Aspect

INTRODUCTION

Today, Russian-Chinese relations have reached an unprecedented level of mutual trust and cooperation. Both countries have established a strong bond in various aspects, including politics, economics, and culture. This elevated level of trust has promoted closer collaboration between the two nations, allowing them to tackle global challenges and make significant progress on various regional and international issues. With the joint efforts of both governments, Russian-Chinese relations have achieved a remarkable level of development that sets an excellent example for other countries seeking to enhance their bilateral partnerships. The deep-rooted political trust between Russia and China has laid a solid foundation for their strategic cooperation, enabling them to strengthen their positions in the global arena and driving the overall stability and prosperity of both nations.

An important part of Russian-Chinese cooperation at the interstate and interregional levels is cooperation in the field of ecology. It becomes especially relevant in the border areas of Russia and China—in particular, in the Far East, where these countries have a common long border. In the border contact zone (Primorsky and Khabarovsk Territories, the Amur Region and the Jewish Autonomous Region), there are continuous processes of interaction in various directions and aspects. In the current geopolitical conditions, environmental cooperation acquires the features of cross-border cooperation.

It should be noted that both Russia and China pay attention to environmental policy issues, but environmental protection issues are not a priority in Russian-Chinese interaction. Much more attention is paid to cooperation in the economic, energy, political, financial, and military spheres. However, close cross-border contacts, joint active development of natural resources and minerals of the Far East puts the environmental sphere in a number of priority and most important areas of cooperation.

The regional aspect of environmental Russian-Chinese cooperation is becoming even more important in the context of the opening of advanced social and economic development zones (ASEZs) in the Far East and the attraction of Chinese investments to the region, as well as under the China's implementation of the global initiative "One Belt, One Road".

The purpose of this study is to characterise the environmental policies of Russia and China and its implementation at the regional level, to consider the cooperation between two countries in the transboundary territories of the Far East in the field of environmental protection, to examine the problems and prospects of environmental cross-border cooperation in the twenty-first century.

A system method was used in this study. This allows me to analyse inter-regional cooperation between China and Russia as one of the components of the overall system of Sino-Russian relations in its historical process. As an element of the system, inter-regional cooperation is influenced by internal and external factors; this chapter focusses on the role of external factor—namely, the role of the environmental pollution on the sustainability of cooperation, its essence and perspectives. Furthermore, a historical-chronological method allows me to consider cooperation through the prism of the history of Russian-Chinese relations and the history of the Russian Far East in the late twentieth and early twenty-first centuries.

Documents Concluded Between Russia and China on Environmental Cooperation and Their Effectiveness

In the late twentieth to early twenty-first centuries, both Russia and China have adopted a number of fundamental documents in the field of environmental protection. The main goals and principles of the Russian environmental policy are enshrined in the following programs: "Fundamentals of the State Environmental Policy of the Russian Federation until 2030" and "Strategy for the Environmental Safety of the Russian Federation for the Period until 2025". The "Strategy" lists the main challenges to environmental security, including "an increase in the consumption of natural resources with a decrease in their reserves, which, against the background of the globalisation of the economy, leads to a struggle for access to natural resources and has a negative impact on the state of national security of the Russian Federation", as well as "Pollution of atmospheric air and water bodies due to the transboundary transfer of pollutants, including toxic and radioactive substances from the territories of other states".[1] In addition, the location on the territory of Russia of environmentally hazardous industries, as well as production and consumption waste by unscrupulous foreign or transnational business structures, the shooting of migratory species of animals, etc. is considered as a threat, but specific states that could pose a danger are not named.

In China, environmental policy became a part of state policy rather late—only in the early 1980s, and its main institutions and regulatory framework were formed in the 1990s, when "green growth" began to be taken into account in GDP and losses from environmental pollution were calculated. This shift in China's approach to environmental policy reflected the country's recognition of the urgent need to address environmental degradation and its commitment to sustainable development. The government's decision to make environmental policy a part of its state policy demonstrated a shift towards a more holistic and long-term perspective on development. By incorporating "green growth" into its GDP calculation and accounting for the economic losses caused by environmental pollution, China signalled its commitment to balancing economic progress with environmental protection. This highlighted a growing understanding that sustainable development was not only essential for the well-being of its citizens and preservation of its ecological resources but also crucial for the country's long-term economic stability and global reputation.

Today, although China actively participates in numerous international environmental forums and conferences, it still has not signed the main international UN Conventions on the Protection of Water Resources of 1992 and 1997, preferring to conclude bilateral agreements. Beijing's environmental policy directly depends on the goals of economic development. The development of clean technologies should not hinder economic growth, so we cannot say that the Chinese consciousness has "environmental friendliness" and the Chinese environmental expertise has the absolute transparency. Without denying significant efforts in the field of reducing harmful industries, in carrying out water protection, soil

---

[1] The strategy of environmental safety of the Russian Federation for the period up to 2025, Russian official Internet portal of legal information, 2017, http://pravo.gov.ru/proxy/ips/?docbody=&firstDoc=1&lastDoc=1&nd=102430636 (06..09.2023).

protection, and other measures, it should be recognised that the features of China's environmental policy are currently part of its image strategy.

Recently, a system of Chinese environmental legislation has been created, normative documents are being developed at the national level (their mandatory element is political guidelines regarding the country's socioeconomic development), and 5-year national plans for environmental protection are being adopted. However, cross-border cooperation and cross-border interaction are practically not reflected in these documents.[2] In the period from 1988 to 1999, Russia and China signed a number of bilateral agreements on cooperation in the field of environmental protection (1994), fisheries (1988), forest fire protection (1995), protection, regulation and reproduction of living water resources (1994), protection of the Amur tiger (1996), aquatic bioresources (1997), about the joint economic use of individual islands and adjacent waters of border rivers (1999). In 2001, the Agreement on Good Neighborliness, Friendship and Cooperation between the Russian Federation and the People's Republic of China was signed (July 16, 2001).

The Art. 19 of this treaty declared cooperation between the two countries in the "fair rational use of border watercourses, living resources in the North Pacific Ocean and the basins of border rivers" and the need for "joint efforts to protect rare species of flora and fauna and natural ecosystems in border areas". This document is, of course, directly related to the situation on the Amur River, along which the border between Russia and China in the Far East passes. The Amur River is the largest transboundary river in Eurasia, in the basin of which there are six constituent entities of the Russian Federation, three provinces of the People's Republic of China, three aimags[3] of the Mongolian People's Republic, and a small part of the DPRK. According to Art. 1 of the Convention on the Protection and Use of Transboundary Watercourses and International Lakes (Helsinki, 1992), "transboundary waters are any surface or groundwaters that signify whether they cross or are located within boundaries between two or more states".

In the communiqué, memorandums and declarations signed by the Russian Federation with the PRC in 2003–2007, it was determined that

---

[2] Kondratenko, Galina. "Problems and opportunities of cross-border cooperation between China and Russia in the field of ecology". Izvestia of the Eastern Institute, sy. 2 (2017): 42–45.

[3] The first-level territorial-administrative unit in Mongolia.

the parties "on the basis of joint environmental monitoring of trans-boundary objects will develop joint measures to improve the state of the Amur and provide for the development of an intergovernmental agreement on cooperation in the field of protection and rational use of transboundary waters".

Separately, the parties spoke about the obligations of the parties to ensure the environmental safety of the Amur. However, in practice, it turned out that these documents are of a declarative nature, they do not elaborate the mechanisms and legal foundations for a bilateral settlement of transboundary problems, and there are no obligations to compensate for damage in the event of transboundary pollution.

ENVIRONMENTAL CATASTROPHES AND ACCIDENTS IN THE TRANSBOUNDARY ZONE OF RUSSIA AND CHINA

These moments were fully exposed during the large-scale pollution of the Amur River in November 2005, caused by an accident at a petrochemical plant in Jilin, China, where several reactors for the synthesis of nitroben-zene exploded. More than 100 tons of nitrobenzene, benzene, aniline, and toluene were poured into the Songhua River. The ecological well-being of both the Songhua River itself, on which the cities of Harbin and Jilin are located and of the ecosystems of the transboundary Amur River for more than a thousand kilometre from the village of Nizhnelenin-skoe (20 km below the confluence of the Songhua with the Amur) to its mouth (Nikolaevsk-on-Amur town) was endangered. Large cities located in this area got into the pollution zone: Khabarovsk, Komsomolsk-on-Amur, Amursk, as well as the settlements along the Amur. Because of this accident, in Harbin, one of the largest and coldest Chinese cities, the water supply was cut for four days (Harbin is supplied with water from the Songhua).[4,5]

This pollution, which had far-reaching consequences both for the water area of the river and for the population, from the point of view of inter-national law, refers to grave offences and is the basis for the emergence

[4] Nosova, Svetlana. "Russia-China: Legal Regulation of Environmental Relations in the Amur River Basin". Power and Management in the East of Russia, sy. 3 (2007): 135–136.

[5] Berdnikov, Nikolai. "Pollution of Amur River in connection with an accident at a chemical plant in Jilin (PRC)", November 13, 2005. Mining information and analytical bulletin, sy. 12 (2007): 297–298.

of international legal responsibility of the subject of international law. However, the current Russian-Chinese Agreement (1994) "On cooperation in the field of protection, regulation and reproduction of living water resources in the border waters of the Amur and Ussuri" established the obligations of states only to control the state of water bodies and take measures to prevention of pollution and elimination of their consequences, without fixing the legal consequences of their violations.[6] In April 2006, at a plant near Harbin another accident was occurred (and again tens of tons of chemicals fell into the river). In August of the same year, another accident occurred at a factory in the Jilin city accident was occurred, as a result of which industrial waste containing benzene compounds, including aniline, got into the tributary of the Songhua River, the Manniu River.

The incidents that took place forced us to seriously talk about the environmental threat from China, although specialists drew attention to the constant pollution of transboundary and border water bodies of the Far East in 1997–1998. Nikolai Efimov, the coordinator of the Amur programme of the Far Eastern branch of WWF (World Wildlife Fund) Russia, describes the situation that developed in the winter of 1997: "Water from taps and river fish began to be given off with carbolic acid, cases of poisoning of domestic animals and even people were registered. In some districts of the region, the authorities officially prohibited the use of water and fish from the Amur. Studies of water samples have revealed a record concentration (906 MPC) of the sum of polycyclic aromatic compounds of the phenol series, which marked the beginning of the 'phenolic' stage of the study of the Amur River". In the 2000s, specialists of the laboratory of medical ecology conducted a large-scale examination of the health status of residents of coastal villages in the region. First, samples of water, fish, river algae, and silt were taken throughout the Lower Amur River. It turned out that the main polluter of the river is oil-refined products. In the villages of the Lower Amur, doctors examined about a thousand people, as a result, in 73% of adults and 80% of children under 14 years of age, liver dysfunction was detected.[7]

---

[6] Nosova, "Russia-China: Legal Regulation", 136.

[7] China poisoning Amur River with thousands of tons of chemicals, Version, 30 August 2010, https://versia.ru/kitaj-travit-amur-tysyachami-tonn-ximikatov (16.08.2023).

Soon after these events, more than 200 sewage treatment plants were built on the Chinese coast. The Chinese have moved several dozen industrial enterprises away from the coast and, together with the Russian side, are regularly monitoring the waters of the Amur River. Environmentalists note that these measures had a positive effect on the quality of water in the Russian territory. The large-scale flood of 2013 also helped in the cleansing of the Amur River—a stream of water washed the bottom sediments from the chemicals accumulated in them.[8]

The Ecological Situation in the Transboundary Zone of Russia and China Today

Today, the ecological situation in the Russian Far East cannot be called favourable. The volume of emissions of pollutants from stationary sources in the Amur and Irkutsk regions, the Republic of Buryatia is increasing. In the Jewish Autonomous Region, a significant role is played by the production of cement, which is among the most polluting the atmosphere. The amount of pollutant emissions is primarily influenced by the production of electrical and thermal energy, which in most regions of the Far East is based on solid fuels. In almost all regions, an increase in the volume of industrial waste was recorded, which is associated with an increase in the scale of mining. Chinese enterprises in the Far East are allowed to mine gold, which is often carried out with serious violations of environmental standards. Illegal deforestation and large-scale smuggling of timber continues into China (as China has a deforestation moratorium). In China, rapid economic growth in 2003–2010 led to a doubling of GDP but it was accompanied by a significant deterioration of the environment, an increase in wastewater discharges, emissions of pollutants into the atmosphere, and the formation of industrial waste by 1.3, 2.4, and 2.6 times.[9] The close proximity of the Far Eastern territories with China has a direct negative impact on the environment. The regions of the Far East bordering on China are regularly smogged due to the burning of grass on the territory of the PRC, which is accompanied by an increased concentration of CO_2 on Russian territory.

[8] Amur River became cleaner thanks to the Chinese, HabInfo, 4 July 2018, https://habinfo.ru/sostoyanie-amur/ (20.08.2023).

[9] Zabelina, Irina et al. "Economic development and negative impact on the environment in the regions of cross-border interaction", ECO, sy. 8 (2016): 69, 78.

In addition to illegal deforestation and river pollution, the proximity of China creates serious environmental problems for the Russian Far East, such as illegal trade in parts and derivatives of rare animals and rare plants and illegal mining of marine biological resources and pollution of the sea. Russian biological resources attract the Chinese consumer, since in China such natural products have practically disappeared already. Both the Chinese citizens themselves and the Russians are engaged in poaching animals and collecting plants by order of Chinese dealers. Parts of a tiger, leopard, bear, a stream of musk deer, ginseng, aralia, tree frogs, trepang, scallop, etc. are illegally exported. Some of these species (tiger, leopard, ginseng) are classified as "rare and endangered" and are protected by state.

The extraction of others is permitted, but often they are obtained in such barbaric ways and in such large quantities that their existence may also be threatened soon. For example, in order to get the musk gland of a male musk deer ("a stream of musk deer"), they put loops on the animals, into which, in addition to adult males, females and young animals fall. Up to 60 thousand animals die a year. In order to collect tree frogs, poison is poured into the ponds, as a result of which all the fauna that lives there perishes. Young trepang is caught, one-two-year-old roots of ginseng are dug, etc.[10]

CONCLUSION

As a conclusion, we should note that, today, we have to admit with regret that the existing cooperation mechanisms and regulatory documents are not working and do not fully solve the environmental problems facing the border regions of the two countries. The Far East retains its focus on raw materials, and most of the projects planned in the state programmes of the Russian Federation are aimed at the extraction and primary processing of natural raw materials—this will increase the already significant environmental load in the border regions of Eastern Russia. Scenarios of regional socioeconomic development are not correlated with specific programme documents in the field of cross-border cooperation.[11] It seems that the

---

[10] Tysyachnyuk, Michail et al. "Networks of international environmental organizations on the Russian-Chinese border: problems and prospects". Journal of Sociology and Social Anthropology, sy. 9 (2006): 48–49.

[11] Zabelina, Irina et al. "Economic development", 79–80.

problems of coordination in the field of ecology lie in several planes. Firstly, in Russia there is a mismatch between the centre and the regions. Environmental action plans adopted by the regions are poorly coordinated with national projects, as well as with the plans of neighbouring regions (despite the presence of a common unifying ecosystem); they are often filled with general phrases and vague wording. Secondly, the regulatory framework of the Russian Federation in the environmental sphere is poorly developed, many key documents (the Environmental Code, the Law on the Protection of the Amur, etc.) have not yet been adopted, environmental problems are not a priority in the minds of the authorities. Thirdly, there is a significant underfunding of regional environmental programmes (in most Russian regions, environmental expenditures do not exceed 1% of the gross regional product (GRP), as a result—the solution of environmental problems based "on the residual principle". Fourth, the population is massively unformed "Ecological" consciousness. As already noted, this is also typical for China, which creates additional difficulties in transboundary environmental cooperation in the Far East, where there is an almost complete absence of industries that provide comprehensive and waste-free processing of raw materials. The creation of ASEZs and border economic zones with special regimes, where the environmental factor is fully taken into account, is still an unattainable dream. In the existing model of Russian-Chinese cross-border interaction, the main role is played by projects in the border area, but within the cross-border territory "it is difficult to find the unity of historical their ethnic and cultural traditions".[12]

Despite the mentioned problems, transboundary environmental cooperation is being carried out in the Far Eastern border territories, and processes of inter-regional interaction are underway. Joint projects are being implemented to monitor river pollution, preserve Red Book cranes, combat epidemiological diseases, etc. Closer attention to environmental problems at the interstate level, strengthening the initiatives of local regional authorities of border areas in the implementation of environmental projects, attracting new technologies in this area—all this will open up new opportunities for environmental cooperation between the two countries and will contribute to deepening and expanding transboundary interaction.

---

[12] Kondratenko, "Problems and opportunities", 41.

REFERENCES

Amur River became cleaner thanks to the Chinese, HabInfo, 4 July 2018, https://habinfo.ru/sostoyanie-amur/ (20.08.2023).

Berdnikov, N. (2007). Pollution of Amur River in connection with an accident at a chemical plant in Jilin (PRC), November 13, 2005. *Mining Information and Analytical Bulletin, sy., 12,* 297–303.

China poisoning Amur River with thousands of tons of chemicals, Version, 30 August 2010, https://versia.ru/kitaj-travit-amur-tysyachami-tonn-ximikatov (16.08.2023).

Kondratenko, G. (2017). Problems and opportunities of cross-border cooperation between China and Russia in the field of ecology. *Izvestia of the Eastern Institute, sy., 2,* 40–58.

Nosova, S. (2007). Russia-China: Legal regulation of environmental relations in the Amur River Basin. *Power and Management in the East of Russia, sy., 3,* 133–140.

The strategy of environmental safety of the Russian Federation for the period up to 2025, "Russian official Internet portal of legal information", 2017, http://pravo.gov.ru/proxy/ips/?docbody=&firstDoc=1&lastDoc=1&nd=102430636 (06.09.2023).

Tysyachnyuk, M., Pchelkina, S., & San, Y. (2006). Networks of international environmental organizations on the Russian-Chinese border: Problems and prospects. *Journal of Sociology and Social Anthropology.* sy. 9, 42–80.

Zabelina, I., & Klevakina, E. (2016). Economic development and negative impact on the environment in the regions of cross-border interaction. *ECO, sy., 8,* 67–82.

CHAPTER 8

Global South and the Resistance to Pan-Securitisation of Environmental Issues: A Case Study of India and China

Ashmita Rana℗ and Justin Joseph

Abstract The Global North-led initiatives to achieve "pan-securitisation" of environmental issues have often been met with vehement resistance from the Global South. This paper explores the foundations, motivations, and other nuances of this resistance, with a particular focus on India and China. It uses discourse analysis to examine how they use the international fora to counter the West's narrative of the intrinsic link between environment and international security. Building on this, it explores the securitisation process beyond its one-way, linear template. Finally, the paper uses the inferences from the cases to test the relevance and utility

A. Rana (✉)
School of International Studies (SIS), Jawaharlal Nehru University (JNU), New Delhi, India
e-mail: ashmit93_ise@jnu.ac.in

J. Joseph
Centre for Climate Change and Environment (CCCE), Vellore Institute of Technology (VIT) Chennai Campus, Tamil Nadu, Chennai 600127, India

© The Author(s), under exclusive license to Springer Nature Singapore Pte Ltd. 2024
J. T. Karackattu et al. (eds.), *Environmental Securitisation in India and China*, https://doi.org/10.1007/978-981-97-9160-6_8

of the Copenhagen School itself by assessing the recent scholarship that has been critical of the school for its civilisationism and methodological biases. In sum, the paper urges a rethinking of an uncritical application of the Copenhagen School's securitisation framework to understand nuances of securitising moves at the international level that involve complex power dynamics.

Keywords Pan-securitisation · Environmental Issues · Global South · India · China · Copenhagen School

INTRODUCTION

Securitisation theory has greatly impacted the way matters of national and international security are understood in International Relations (IR). The theory unpacks an issue's "social construction" from "normal politics" to a high-priority security one that needs urgent attention. By employing an alarmist rhetorical structure ("speech act"), the "securitising actor" (actors who try to move an issue beyond "normal politics") attempts to convince the target "audience" to take extraordinary measures in order to address the issue, which if successful, securitises the "referent object" (object under threat that needs protection) (Buzan et al., 1998). There-fore, the theory sheds light on how matters of "security" are not entirely objective but are securitised through collective practices. The central objective of this chapter is to apply the securitisation theory to appraise the emergence of the notion of "environmental security"—an emerging concept spearheaded by some influential countries of the Global North but resisted by a majority of the Global South.

The chapter begins by tracing the origins of the efforts towards the pan-securitisation of environmental issues, understanding the actors behind it and their discursive practices at the global fora. It then moves to analyse the role of the "audience", highlighting their agency in resisting the securitising moves. The case studies of India and China are used to illustrate this. India and China are the two most important emerging powerhouses in Asia. With their rising material power, their assertion in the ideational realm has also visibly increased. Their cases provide a window to analyse how the dominant discourse is challenged, and secu-ritisation is resisted at the international level. Finally, the inferences from the case studies are used to assess the relevance of the securitisation theory of the Copenhagen School in understanding the Global South's role amid rising narratives around "environmental insecurity".

PAN-SECURITISATION OF ENVIRONMENTAL ISSUES: FOR WHOM AND BY WHOM?

Securitisation of an issue, no matter how objectively urgent or minacious, is never wholly apolitical. Buzan, Wæver and Wilde have famously argued, "...it is always a political choice to securitise or to accept a securitisation" (Buzan et al., 1998, p. 29). International anarchy and asymmetric global power relations intertwine in multiple ways and at multiple levels to shape the securitisation narrative around environmental issues. While environmental issues had started gaining momentum in the global discourse in the 1970s, it was in the Brundtland Report of 1987 that the phrase "environmental insecurity" was used for the first time (Gordeeva, 2022, p. 9). The report talked about a causal and sometimes even "catalytic" link between environmental stress and major conflicts (WCED Report, 1987). Interestingly, this causal link had an intervening variable—inept political processes. The report argued that environmental stress, when unable to be handled by concerned political processes, can add up to the "web of causality" related to a conflict (WCED Report, 1987). The report denounced any military solution to environmental insecurity but argued that interdependence in the domain of economics, environment and insecurity had "fundamentally modified" the idea of national sovereignty (WCED Report, 1987). Following the Brundtland Report, there has been a steady growth in the global discourse on the interdependence between environment and international peace and security across various international fora.

The United Nations Security Council (UNSC) has been one of the main platforms where the push for and against pan-securitisation of environmental issues has been the most prominent. Being the most powerful international body and the one in which all the major powers have a stake, the UNSC has been at the forefront of such deliberations. In 2007, the UNSC held its first-ever debate on the "impact of climate change on peace and security". Countries like the United States and the United Kingdom rallied for a greater role for the Council in climate change as a collective security issue (Department of Public Information, UNSC, 2007). While dismissing apprehensions of a few countries from the Global South that the UNSC would be encroaching on the roles and responsibilities of other organisations, these countries emphasised the security implications of environmental issues like border disputes, conflict over scarce resources, mass migrations, etc. The Council, in particular, outlined that climate's implications on peace and security were especially pronounced for vulnerable regions with "pre-existing conflict, poverty and unequal access to resources, weak institutions, food insecurity and incidence of diseases such as HIV/AIDS" (Department of Public Information, UNSC, 2007). Similarly, several thematic debates and discussions on the issue have been held in the UNSC over the years.

A recent UNSC Ministerial Open Debate on the environment and security nexus is an important case to analyse as it offers a general insight into how the Global North has been shaping the discourse on pan-securitisation of environmental issues at global fora. In the debate, the head of the European Union (EU) delegation began their address by stressing the vulnerable regions and groups to be impacted by the security implications of climate change simultaneously (Press and Information Team, EU Delegation to the UN in NY, 2023). They then proceeded to argue for an increased role of the UNSC at the behest of the affected countries and populations. The EU delegation called for a greater systematic engagement between the UN Climate and Peace and Security Mechanism, deployment of environmental advisors to EU-UN civilian and military missions, increased coordination with OSCE (Organisation for Security and Cooperation in Europe) and NATO (North Atlantic Treaty Organisation), among others (Press and Information Team, Delegation to the UN in NY, 2023). In sum, the debate, like most debates on the theme, began by arguing in favour of an intrinsic link between environmental issues and international security and then highlighted the plight of the most vulnerable regions (Global South) having, inter alia, weak governance institutions or inept political processes, before concluding with the need to develop synergies between environmental organisations and other security organisations, including even military ones like NATO. Therefore, a pattern would be visible if one were to unpack the discourse on the environment-security nexus led by the Global North. First, the risk and urgency of environmental issues are laced with the larger discourse on international peace and security to advocate for a greater role of the UNSC. Second, the rhetoric that accompanies the discursive practices for the pan-securitisation of environmental issues engages with the Global South, highlights their plight to evoke a sense of urgency and contend for an expansion in the UNSC's mandate. Third, the regions and countries named under the vulnerable category (from the Global South) are depicted as inept in handling environmental stresses because of a number of problems, but most notably due to "weak" political/governance institutions. Fourth, while military solutions to environmental insecurity are either denounced or not mentioned at all, engagement with military organisations like NATO is encouraged. There have even been talks on the UNSC's possible role in "green peace-keeping" in the past, raising speculations on the nature of these missions (Goldenberg, 2011).

Thus, if the pan-securitisation discourse on the environment (*referent object*) is analysed, we can infer answers to two critical questions involved in any securitisation act—securitisation by whom (*securitising actor*)? And securitisation for whom (*audience*)? Clearly, the pan-securitisation of environmental issues is spearheaded by major countries of the Global North and the international institutions within their influence. The Global South appears to be the audience for the former's securitising move within a specific iterative rhetorical structure.

Resistance by the Global South: Not a Passive "Audience"

Among the major critiques of the Copenhagen School is its conceptual-isation of the audience as a passive actor lacking agency of their own. Scholars have argued that theorising the securitisation audience as an active agent that does not simply accept or reject the threat narrative of the securitisation move but interacts with the securitising actor in many ways is more reflective of the reality (Côté, 2016; Floyd, 2020; Stępka, 2022). Building on this argument, this chapter contends that the pan-securitisation of environmental issues discourse has been contested by many countries of the Global South at international fora, especially at the UNSC debates. This contestation is based on a number of apprehensions and scepticism that the Global South is confronted with. Firstly, integrating environmental issues with questions of international peace, stability, and security exposes states to the possibility of military actions like "ecological" interventions or sanctions (Gordeeva, 2022, p. 7).

While the possible use of military measures has been dismissed at platforms like the UNSC, there have been mentions of increased coordination with military alliances like NATO at the same time. Secondly, it is argued that the tendency to focus on environmental threats to peace originating from countries with "weak" governance systems and institutions is rooted in racist and colonial stereotypes against the Global South (Cusato, 2022, p. 230). The narrative is such that it shifts the focus from the major emitters who have a long history of environmental degradation to the "weak" countries who cannot handle environmental stresses, thereby becoming the origin of a "security threat". Thirdly, even if coercive interventions in environment "rogue" or "failed" states were to be ruled out, the possibility of compromise with the sovereignty of some vulnerable countries in the name of collective security lingers. The principle behind this strikingly mirrors the R2P (Responsibility to Protect) doctrine, which justifies an interventionist discourse using the perception that the Global South is mainly dysfunctional and lacks "good governance" as defined by Western benchmarks (Hindawi, 2021). Fourthly, the discourse on environmental security is often perceived by critics as a discourse about the security and economic interests of the northern countries (Trombetta, 2009, p. 586). The "global" discourse is not representative of a universal human interest; rather, it reflects the parochial interest of the powerful who have the reach and control to globalise their interests (Shiva, 1994).

The aforementioned principles have shaped the counter-narrative to the Global North's pan-securitisation of environmental issues. The countries from the Global South, who have been at the forefront of this opposition, have argued that the largest emitters of greenhouse gases have veto power in the UNSC. This is ironic as the primarily guilty states possess such a massive advantage while the main victims remain inadequately represented (Cusato, 2022, p. 224). The following section will examine the case of two main voices against the pan-securitisation of environmental issues led by the West. India and China, with their rising economy and consequent political clout in world affairs, have been two of the biggest challengers to the dominant West-centric discourse.

India

India has been proactive in taking the lead in voicing environment-related concerns of the Global South for a long time. This section will mainly be focussing on India's response to the discourse on the pan-securitisation of environmental issues. The very first time that the climate-security nexus was discussed in the UNSC, India denounced the possibility of any meaningful link between a political argument and an objective scientific phenomenon (Department of Public Information, UNSC, 2007). On the other hand, it argued that a graver threat was from potential conflicts arising from inadequate resources, leading to underdevelopment, poverty and competition for energy (Department of Public Information, UNSC, 2007). India also questioned the UNSC's mandate on the matter and argued in favour of fora like the UNFCCC (United Nations Framework Convention on Climate Change) as being more appropriate for environment-related deliberations (Department of Public Information, UNSC, 2007). What was notable in India's arguments was that it sought to shift the attention back to the developed countries who were somehow downplaying their original role and responsibility in the crisis by talking about peace and security. By portraying the "weak" and "vulnerable" countries as a potential source of conflict, the dominant discourse was displacing responsibility and perpetuating problematic assumptions about the Global South (Cusato, 2022). Via its response, India sought to urge the developed countries to take greater responsibility for reducing their emissions and energy consumption levels. Also, by advocating how harsh greenhouse gas mitigation targets would impact the development levels of other countries and increase their insecurity, India highlighted the often-sidelined circumstances of the Global South. In 2021, during its tenure as a non-permanent member, India voted against a draft resolution on climate change in the UNSC. While explaining its stance, India argued that viewing conflicts through the prism of climate change comes at the cost of over-simplification and is thus misleading (Amb. T.S. Tirumurti, 2021). The Indian representative did not mince any words while calling out developed countries for evading responsibility, diverting attention from their commitments and side-tracking the real issues under the "guise of security" (Amb. T.S. Tirumurti, 2021).

Apart from challenging the West-led discourse, India has also led initiatives of its own on environmental issues at both global and regional levels. The most notable of these is the leadership it has demonstrated in the setting up of two international organisations—the International Solar Alliance (ISA) and the Coalition for Disaster Resilient Infrastructure (CDRI). It is insightful to look at the mandate, agenda and general flow of discussions in these organisations. For instance, ISA, an alliance of solar resource-rich countries, was launched by India in 2015. During the launch and initial phase of the alliance's mobilisation for membership, India used its domestic achievements in the domain of environment and climate to cultivate for itself an image of a responsible power that was part of the "solution" and willing to take on greater responsibility than its "fair share" (PIB Delhi, 2023a). Moreover, it also emphasised "climate justice" as a key pillar of the ISA (PIB Delhi, 2018). Simply put, climate justice is a concept that, inter alia, stresses the unfair paradox that developing and underdeveloped countries face. While the historically big emitters have the best resources to cope with the consequences of environmental degradation, the poor and marginalised face the most extreme outcomes of the issue (Macquarie, 2022). Clearly, India's discourse is not centred around alarmist rhetoric propagating pan-securitisation. Instead, it is reminiscent of a South-South solidarity sentiment. The initiatives undertaken by the ISA to date have been directed towards enhancing the beneficiary countries' agency by establishing training centres, conducting capacity-building programmes, etc. India has also used ISA as a platform to urge the developed countries to fulfil their green funds commitments (PIB Delhi, 2023b). Similarly, CDRI is an initiative in which vulnerable countries are assisted as per their specific requirements for infrastructure resilient to environmental hazards. A number of other initiatives have been spearheaded by India within existing organisations as well that aim at delivering country-specific solutions in the Global South. These initiatives are not conceptualised as independent of the major powers but aim to urge them to take greater responsibility towards the Global South by engaging with the latter's often unique requirements.

Thus, a brief overview of India's engagement with the discourse on pan-securitisation of environmental issues reveals two major approaches: firstly, it openly opposes the West's attempt to include environmental challenges in the existing international peace and security paradigm; secondly, it tries to engage with the Global North at alternative platforms while shaping norms for their ethical responsibilities towards the Global South.

China

The Chinese case is similar to India's in terms of its stance towards the issue, but there are some key divergences in Beijing's approach from New Delhi's. China has a history of either sometimes abstaining or mostly vetoing any resolution in the UNSC that intends to integrate environment-related hazards with security risks that are associated with existing conflict prevention and resolution frameworks. In the first-ever meeting of the UNSC on the issue in 2007, China argued that the UNSC was not a platform where extensive participation by all countries that would result in widely acceptable proposals was possible (Department of Public Information, UNSC, 2007).

China has consistently expressed strong criticism during UNSC debates at the prospect of assigning climate change-related mandates to peace-keeping missions (UNSC Report, 2022). It is noteworthy that China has often disagreed with the human rights language used in such resolutions. Human rights have been one of the ideological arenas where China has fiercely challenged the US. The Chinese have quite a different opinion on human rights from the Anglo-Saxon understanding of the concept. The different historical, ideological, political and social conditions have led to competing outlooks on human rights between the two (Qi, 2005, p. 105). China regards sovereignty as sacrosanct and has been critical of the West for using human rights as a guise for violating it. Moreover, the CCP (Chinese Communist Party) aspires to strengthen *huayuquan*, which means its discourse power to tell the Chinese side of the story on the international stage (Wang, 2015). In a way, the competing discourse on human rights superimposes with the one against pan-securitisation of environmental issues to shape Beijing's diplomatic standing on the topic. China's scepticism about the possibility of the West using humanitarian grounds for intervening in other countries that pose an environment-related security risk shapes its stand on the issue to a great degree. Thus, it has been against the UNSC expanding its mandate concerning environmental issues, on one occasion accusing the UNSC of making the issue a "political show" (Xinhua, 2021).

In addition, China has maintained a strong position that environmental issues like climate change should be seen as questions of (sustainable) development rather than security (Scott, 2012, p. 227). It has been a passionate advocate for CBDR (common but differentiated responsibilities), a principle that questions the applicability of the same standards for the most advanced countries and the developing countries when it means unwarranted social cost for the latter (Diez, 2014). Calling out for the developed countries to take responsibility for their past actions, China, much like India, has promoted South-South cooperation in this regard (Xinhua, 2021).

Like India, China has also led initiatives to counter environmental issues. However, there are certain unique attributes of the Chinese case. First, it has preferred to engage bilaterally with countries more than in multilateral forums with American influence. For instance, as of June 2023, it had signed 46 MoUs (memorandums of understanding) on climate change cooperation with 39 developing countries (Zhou & Daojiong, 2023). Second, it has integrated environmental assistance to the Global South in its existing economic development projects like the BRI (Belt and Road Initiative). The BRI International Alliance for Green Development, the BRI Ecological and Environmental Protection Big Data Service Platform and the Technology Transfer South–South Cooperation Center are some examples of projects intended to engage with the Global South on the environment within the BRI framework (Zhou & Daojiong, 2023). Third, while India has spearheaded new institutions and sought global participation in them, China has focussed more on regional institutions without the membership of great powers like the US. A few examples include the China-Arab States Cooperation Forum and the Forum on China-Africa Cooperation, which, among other things, deal with green finance.

In sum, China has opposed the pan-securitisation of environmental issues by the UNSC by arguing for the primacy of sustainable development. Moreover, it has assumed leadership in the Global South using bilateral or regional institutions. Green finance or climate aid is an area where it has been more active. Interestingly, it is going a step ahead by not only trying to reshape the existing modalities for delivering (environmental) aid but also by creating some new ones (Freeman, 2020). Thus, in China's case, the clash of discourses on environmental issues is yet another ideational manifestation of its larger geopolitical competition with Washington.

RETHINKING COPENHAGEN SCHOOL FRAMEWORK FOR UNDERSTANDING THE GLOBAL SOUTH'S RESISTANCE

The Copenhagen School's securitisation theory has been quite significant and influential for International Relations scholarship. However, it has come under a fair share of criticism as well, primarily from the non-West. Scholars have argued that the securitisation framework of the Copenhagen School is based on a European understanding of concepts, assumes them to be universal and is often unsuited to non-Western contexts (Holbraad & Pedersen, 2012; Wilkinson, 2007). More recently, harsher criticism has been levelled against the school, calling it out for "civilisationism", "racism" and a "methodologically and normatively white framework" (Howell & Richter-Montpetit, 2019). It has been argued that the conceptualisation of "normal politics" is based on an inherent idea that Europe is the apex of civilized desecuritisation, while speech act theory can be used to locate "progress" towards normal politics and thus towards becoming more like Europe (Howell & Richter-Montpetit, 2019, p. 3). This chapter deals with securitisation at the international level and not within a country's domestic environment. Thus, a discussion of all of these criticisms is beyond the scope of this chapter. However, building on these broad lines of questioning, the insights from case studies in this chapter can be of some value to test the applicability of the Copenhagen School's securitisation framework in the Global South.

Buzan, Wæver and Wilde, the most prominent thinkers of the Copenhagen School, had cautioned that securitising moves such as "environmental security" have negative repercussions of applying a security mindset against "possible advantages of focus, attention and mobilisation" (Buzan et al., 1998, p. 29). In their book, *Security: A New Framework for Analysis*, Buzan et al. attempt to present the essence of different discursive practices around the securitisation of environmental issues in a comprehensive manner yet fall short of capturing the voice from the Global South. For instance, they only mention it as one of the many debates. It is reduced to a "powerful agenda for the periphery against the centre" and gets linked to the NIEO (New International Economic Order) North–South debate, which argues how the dominance of Western values legitimises inequities in different sectors, mirroring the structures of poverty and affluence that also lead to environmental degradation among other things (Buzan et al., 1998, p. 82). In addition, they mention how the West's agenda is oriented on the role of population while the South's emphasises economic issues (Buzan et al., 1998, p. 75). When the authors try to point to the "basic logic" of environmental security, they refer to the question of whether there are "civilised" ways to overcome the problems created by "civilisation" itself (Buzan et al., 1998, pp. 81, 82). This inadvertently makes them fall into the trap of a Eurocentric bias. What is referred to as "civilisation" is actually the West's idea of how a society should be, which leads to certain types of consumption patterns that have proved to be highly unsustainable today. The dominance of the Global North in the forums, which set the agenda for environmental issues today and press for "civilised" measures without addressing the historical wrongs or the present obligations, cannot be seen as merely a securitizing move. As has been argued by India and

China, it is a move to displace responsibility to the developing countries for their population spurts and industrialisation when maintaining the current lifestyle for an individual in the developed countries is, if not more, equally harmful to the environment. Moreover, there have been emerging voices arguing against the neocolonial nature of "false" solutions like carbon trading and carbon offsets proposed by the West as they legitimate affluent countries' emissions (Cusato, 2022; Wang, 2021).

The Copenhagen School has only looked at desecuritisation from an ethnocentric lens, suggesting Europe's "civilised" desecuritisation as a "moral imperative" while assuming many Global South states to be prone to "oversecuritisation" because of them being "weak" or "failed" (Howell & Richter-Montpetit, 2020). As argued in this chapter, similar stereotypes for most of the Global South have been used in the West's discourses, arguing for environmental issues to be included in the international security paradigm. Buzan et al., while discussing environmental security, have commented that "weak" or less "resilient" states are more likely to have political insecurities arising from environmental problems (Buzan et al., 1998, p. 88). There seems to be a paradox here in that while (over) securitisation is problematised, when it comes to pan-securitisation of global issues led by the Global North, the same reasons which are used to argue against the former are used to support the latter. It becomes worth wondering what comprises "over" in securitisation and who decides it.

The Global South, especially emerging challengers to Western dominance like the BRICS countries, have contested the West's securitising moves. Even countries with fewer resources and lesser political clout but facing existential threats like the SIDS (Small Island Developing States) have been vocal about only UNFCCC being the appropriate forum for discussing the solution to their extraordinarily urgent problem (AOSIS Statement, 2023). Unlike what Buzan et al. believed when they said that developing countries being victims who lack resources are likely to be support actors (Buzan et al., 1998, p. 78), defying the "issue-specific nature" of positions, the Global South has shown strong resistance against the securitising moves of the Global North. Moreover, if one focusses on the Copenhagen School's conceptualisation of the speech act as a simple linear process, one loses out on the nuances of the process involved due to sole emphasis on the outcome of successful securitisation (Wilkinson, 2007, p. 8). In order to fully appreciate how the Global South has been able to prevent the pan-securitisation of environmental issues, we need to immerse ourselves in the foundations, motivations and responses of their resistance. At the same time, their struggle to posit an alternative discourse on environmental issues can only be fully understood by minimising the biases in the Copenhagen School, which privileges Western ideas and values.

CONCLUSION

This chapter has attempted to illuminate how the discursive practices of some countries of the Global North are intended to securitise the environment to a degree that involves linking it to matters of international peace and security. These discourses are rooted in some stereotypes about the Global South and tend to displace the responsibility for environmental degradation and equitable commitments to tackle the same. However, the Global South, particularly emerging countries like India and China, has strongly resisted any securitising move meant to link environmental issues with international security. The motivations for and nuances of this resistance are reflective of the agency of the Global South and their fight to question the dominance of Eurocentric narratives even in issues that concern the whole of humanity. Additionally, the linear and simplified framework of securitisation of the Copenhagen School is inadequate to fully understand the resistance of the Global South. The internal biases and the privileging of Western ideas in the theory hinder it from being a comprehensive tool for understanding the emergence of the issue of environmental insecurity and the Global South's resistance to it.

REFERENCES

AOSIS. (2023, February 14). *AOSIS statement: United Nations security council open debate on threats to international peace and security: Sea-level rise – Implications for international peace and security.* Alliance of Small Island States. Retrieved February 15, 2024, from https://www.aosis.org/aosis-statement-united-nations-security-cou ncil-open-debate-on-threats-to-international-peace-and-security-sea-level-rise-implications-for-international-peace-and-security/

Buzan, B., Wæver, O., & Wilde, J. D. (1998). *Security: A new framework for analysis.* Lynne Rienner Publishers.

Côté, A. (2016). Agents without agency: Assessing the role of the audience in securitization theory. *Security Dialogue, 47*(6), 541–558. https://www.jstor.org/stable/26293812/

Cusato, E. (2022). Of violence and (in)visibility: The securitization of climate change in international law. *London Review of International Law, 10*(2), 203–242. https://doi.org/10.1093/lril/lrac015

Diez, C. (2014, March 13). Policy brief and proposals: Common but differentiated responsibilities. International movement ATD fourth world. Retrieved January 23, 2024, from https://sustainabledevelopment.un.org/getWSDoc.php?id=4086

EU Statement—UN Security Council: Ministerial Debate on Climate Change, Peace and Security (2023, June 13). Retrieved February 7, 2024, from https://www.eeas.europa.eu/delegations/un-new-york/eu-statement-%E2%80%93-un-security-council-ministerial-debate-climate-change-peace-and-security_en?s=63

Floyd, R. (2020). Securitization and the function of functional actors. *Critical Studies on Security, 9*(2), 81–97. https://doi.org/10.1080/21624887.2020.1827590

Freeman, C. P. (2020, September 11). Reading Kindleberger in Beijing: Xi Jinping's China as a provider of global public goods. *The British Journal of Politics and International Relations, 23*(2), 297–318. https://doi.org/10.1177/1369148120941401

Goldenberg, S. (2011, July 20). UN security council to consider climate change peacekeeping. *The Guardian.* Retrieved February 12, 2024, from https://www.theguardian.com/environment/2011/jul/20/un-climate-change-peacekeeping

Gordeeva, E. (2022, July). The securitization of global environmental policy: An argument against. *European Journal for Security Research, 7*(1), 5–20. https://doi.org/10.1007/s41125-022-00083-x

Hindawi, C. P. (2021). Decolonizing the responsibility to protect: On pervasive eurocentrism, southern agency and struggles over universals. *Security Dialogue, 53*(1), 38–56. https://doi.org/10.1177/09670106211027801

Holbraad, M., & Pedersen, M. A. (2012, June 15). Revolutionary securitization: An anthropological extension of securitization theory. *International Theory, 4*(2), 165–197. https://doi.org/10.1017/s1752971912000061

Howell, A., & Richter-Montpetit, M. (2019, August 7). Is securitization theory racist? Civilizationism, methodological whiteness, and antiblack thought in the Copenhagen School. *Security Dialogue, 51*(1), 3–22. https://doi.org/10.1177/0967010619862921

PIB Delhi. (2018, October 2). *PM inaugurates first assembly of the International Solar Alliance.* Press Information Bureau. Retrieved February 17, 2024, from https://pib.gov.in/Pressreleaseshare.aspx?PRID=1548295

PIB Delhi. (2023a, February 2). *Minister of state for environment, forest and climate change, Shri Ashwini Kumar Choubey in a written reply to a question in Rajya Sabha*. Press Information Bureau. Retrieved February 1, 2024, from https://pib.gov.in/PressReleasePage.aspx?PRID=1895857

PIB Delhi. (2023b, October 31). *India hosts the 6th session of the international solar alliance assembly in New Delhi*. Press Information Bureau. Retrieved February 1, 2024, from https://pib.gov.in/PressReleaseIframe Page.aspx?PRID=1973449

Qi, Z. (2005, February). Conflicts over human rights between China and the US. *Human Rights Quarterly, 27*(1), 105–124. https://doi.org/10.1353/hrq.2005.0011

Scott, S. V. (2012). The securitization of climate change in world politics: How close have we come and would full securitization enhance the efficacy of global climate change policy? *Review of European Community and International Environmental Law, 21*(3), 220–230. https://doi.org/10.1111/reel.12008

Security council holds first-ever debate on impact of climate change on peace, security, hearing over 50 speakers (Security Council SC/9000). (2007, April 17). Department of Public Information, UNSC. Retrieved January 28, 2024, from https://press.un.org/en/2007/sc9000.doc.htm

Shiva, V. (1994). Conflicts of global ecology: Environmental activism in a period of Global reach. *Alternatives: Global, Local, Political, 19*(2), 195–207. https://doi.org/10.1177/030437549401900208

Stępka, M. (2022). The Copenhagen school and beyond. A closer look at securitization theory. In *IMISCOE research series* (pp. 17–31). https://doi.org/10.1007/978-3-030-93035-6_2

Tirumurti, Amb. (2021, December 13). *Explanation of vote by ambassador T.S. Tirumurti permanent representative of India to the United Nations*. Retrieved February 5, 2024, from https://pminewyork.gov.in/IndiaatUNSC?id=NDQ1NQ

Trombetta, M. J. (2009). Environmental security and climate change: Analyzing the discourse. *Cambridge Review of International Affairs, 21*(4), 585–602. https://doi.org/10.1080/09557570802452920

UNSC. (2022, December 30). *The UN security council and climate change: Tracking the agenda after the 2021 Veto* (22(4)). UNSC Report. Retrieved January 31, 2024, from https://www.securitycouncilreport.org/atf/cf/%7B6 5BFCF9B-6D27-4E9C-8CD3-CF6E4FF96FF9%7D/unsc_climatechange_2022.pdf

Wang, J. (2021, May 3). *Carbon offsets, a new form of neocolonialism*. Columbia Climate School, Columbia University. Retrieved February 3, 2024, from https://climatesociety.ei.columbia.edu/news/carbon-offsets-new-form-neocolonialism

WCED. Report of the World Commission on Environment and Development: Our Common Future. (1987, March 20). World Commission on

Environment and Development. https://sustainabledevelopment.un.org/con
tent/documents/5987our-common-*future.pdf*

Wilkinson, C. (2007). The Copenhagen School on tour in Kyrgyzstan: Is securitization theory useable outside Europe? *Security Dialogue, 38*(1), 5–25.
https://doi.org/10.1177/0967010607075964

Xinhua. (2021, December 14). Chinese envoy warns against pan-securitization
of climate issues. *Xinhua Net*. Retrieved February 10, 2024, from http://
www.news.cn/english/2021-12/14/c_1310371572.htm

Zhou, J. & Daojiong, Z. (2023, November 28). *Climate finance
and geopolitics: The China–US factor*. Retrieved February 1, 2024,
from https://www.sipri.org/commentary/essay/2023/climate-finance-and-
geopolitics-china-us-factor

PART IV

State, Society and Environmental Policy Processes in India And China

Climate Securitisation in China and India: A Human Security Perspective

Shilpi Ghosh and *Gajendra*

Abstract This research paper discusses the topic of the securitisation of climate change which is a developing issue that affects countries, people, and the environment. Although the risks associated with climate change are widely acknowledged, not enough is being done to reduce insecurity it causes. This paper investigates how China's and India's national security discourses express concerns about climate security. It attempts to assess the extent to which these countries acknowledge climate change as a threat to national security by examining key security documents from 2014 to 2023. In addition to China's defence and environmental policy white papers, the research examines the important policy documents such as the Annual Reports of India's Ministry of Defence and Ministry of

S. Ghosh (✉)
Centre for African Studies, School of International Studies (SIS), Jawaharlal Nehru University (JNU), New Delhi, India
e-mail: shilpi30_isi@jnu.ac.in

Gajendra
Centre for Diaspora Studies, Central University of Gujarat, Gandhinagar, India
e-mail: gajendra@cug.ac.in

J. T. Karackattu et al. (eds.), *Environmental Securitisation in India and China*, https://doi.org/10.1007/978-981-97-9160-6_9

145

Environment, Forests, and Climate Change. Emphasising the existential threat that climate change poses to both countries, the study aims to highlight the need for implementing a human security approach to effectively address the shortcomings in the current measures against it.

Keywords China · Climate securitisation · Human security · India

INTRODUCTION

Although traditional security and power politics have always attracted a large amount of scholarship, discussions and debates nowadays also focus on other security-related topics, such as food, health, environmental, and economic security. The bipolar framework of the Cold War ended with the description of world geopolitics as a struggle between global ideologies, capitalism and communist. As a result, there are now many different types of non-state actors interacting with state power in more dispersed, complicated, and de-territorialised multi-centric concentrations. These include international organisations, NGOs, media, the internet, multinational enterprises, nascent governments, and "failed states." Thus, previously concealed and underutilised issues as well as local and regional tensions were now emerging. This was the first time that the industrialisation, consumerism, liberalisation, and globalisation constituting the dominant paradigm of development at the very expense of nature, was having a significant impact on a number of key global and regional systems. It appeared that this had put the development of scientific and technological, economic, foreign, and security policy in jeopardy.

Thus, it was believed that the traditional definition of national security—which was based on the state and its armed forces, marked by neo-realism and realism—was inadequate to protect the individuals who were vulnerable to threats from non-traditional form of securities. The securitisation theory, in contrast to other mainstream International Relations theory, provides "a new framework of security" that is accommodative of many transformational ideas. The term "securitisation" was first used in 1995 by Ole Wæver of the Copenhagen School in his paper "Securitizations: Taking stock of a research program in Security Studies." He claimed that "Security" is the outcome of a political manoeuvre that elevates the subject above standard politics and transcends the established norms (McDonald, 2008). Thus, securitisation might be considered a more severe form of politicisation. It is the practice of portraying a problem as so critical and urgent that it should be handled decisively by leaders before other matters, shielding it from the typical political backroom wrangling.

Barry Buzan, another important proponent of securitisation theory suggested that the problem of "national security" cannot be understood without referencing to all three levels of analysis (individual, state, and international system), thus taking a multi-sectoral approach (Buzan, 1984). It is not only the political sphere of state that needs to be protected but its social, environmental, and economic spheres are equally important. It then becomes important to identify the type of interactions happening that are shaped by various actors and levels (Buzan, Wæver, and de Wilde, 1998). These actors present at various levels exhibit both interdependence and independence in their actions. Within the environmental sector, for instance, there is a vast array of potential referent objects.[1] These can range from relatively concrete things like habitat types or individual species survival to a much larger-scale issues like maintaining the planetary climate and biosphere within the narrow range that humans have evolved to consider normal over the course of their few thousand years of civilisation. Hence, it is all about how humans relate to the rest of the biosphere and whether that relationship can be maintained without running the risk of catastrophically disrupting the planet's biological legacy, the collapse of civilisation, or both are some of the referent objects. It is incredibly complex how these factors interact with one another. Certain instances

---

[1] The speech-act approach to security requires a distinction among three types of units involved in security analysis.

    i.   Referent objects: things that are seen to be existentially threatened and that have a legitimate claim to survival.
    ii. Securitising actors: actors who securitise issues by declaring something—referent object—existentially threatened.
    iii. Functional actors: actors who affect the dynamics of a sector. Without being the referent object or the actor calling for security on behalf of the referent object, this is an actor who significantly influences decisions in the field of security. A polluting company, for example, can be a central actor in the environmental sector—it is not a referent object and is not trying to securitise environmental issues (quite the contrary).

of existential threat—the survival of species and human civilisation, for example—can be securitised at either the macro or micro extremes.

It is also significant to highlight here that the Copenhagen School believes that "security should be seen as a negative, as a failure to deal with issues of normal politics." Thus, "de-securitisation"[2]—moving topics from the realm of exceptionality into the mainstream of public discourse—is preferred by the Copenhagen School (Buzan, 1983, 1984). Also, the "threat-defence" framework's character is shaped by the security and insecurity components of climate change, which are shaped by this discourse. One main goal of the Copenhagen theory is to show the effects of securitisation. Securitisation of a situation has consequences that are both "internally," where it can obstruct democracy and discourse, and "externally," where it commonly causes conflict, security problems, and escalation. The threat posed by climate change is not limited to state security, despite its securitisation. Additionally, it poses a risk to the stability of the political system, the economy, and society (Buzan, 1998).

Climate change, as one of the most severe concerns of the twenty-first century, is having a growing influence on human security worldwide. As a threat multiplier, it exacerbates resource scarcity while aggravating a country's current social, economic, and environmental issues. Climate change-vulnerable countries are among the world's most politically and economically fragile areas. According to a report published by the US Military Advisory Board, which consists of a group of highly respected retired Admirals and Generals, "global climate change presents a new and very different type of national security challenge (Gordeeva, 2022)." Thus, climate change is more than just rising temperatures. It has the potential to produce many of the changes in natural systems that have contributed to country instability over the millennia. Today, climate change's national security consequences are being integrated into the security and defence agendas of an increasing number of countries.

The reference point for the study will not be climate change iself; rather, it will focus on human security, framing climate change as a constructed threat to human security. Here securitisation attempts can be

---

[2] Wæver, Ole. "Securitization and De-securitization." In *On Security*. Edited by Ronnie Lipschutz, 46–

a.   New York: Columbia University Press, 1995.

identified by the existence of essential elements like an existential threat on individual, state as a security concern, exceptional measures are sought for these referent objects. Two main issues guide this research analysis of the national security documents of India and China: first, does the policy documentation acknowledge the threat posed by climate change, and second, does it further securitise the issue? An overview of the scope of a security agenda issue can be obtained for the study by calculating the proportion of the total number of pages in the policy document that address environmental concerns (Floyd, 2007).

REVIEW OF INDIA'S POLICY DOCUMENTS

The NDA administration, led by the present prime minister, Mr. Narendra Modi, was elected to power in 2014. The Ministry of Defence and Ministry of External Affairs' annual reports from 2014 to 2023 will be analysed for the study. The number of pages or references in a policy document that addresses environmental issues relative to the total number of pages provides a general idea of the issue's importance in a security agenda. The use of term "Climate Change" finds once (the Institute for Defence Studies and Analyses, or IDSA, conducts research on a wide range of modern topics, including water security, energy, and climate change) or no reference at all in the eight reports (2014–2022) that were analysed for the research. Potential environmental issues are covered in fewer than one per cent of the annual defence reports. The reports have always emphasised how India's defence strategy and policies have always sought to create a peaceful environment by tackling the vast array of conventional and non-conventional security issues like cyber security, pandemics, natural disasters, terrorism, transnational crime, and food and energy security problems that the nation faces.

The reports do not explicitly suggest that climate change poses a security risk. The benefit of using such a phrase in the context of securitisation is that it might make the problem a top priority on a security agenda and start the process of taking the appropriate steps to address the threat posed by climate change. Every Pollution Control Board regulation has been scrupulously adhered to (MOD Annual Report, 2014–2015; 2017–2018).

The National Institute of Solar Energy (NISE) hosted the first Tri-Services "Non-Conventional & Renewable Energy" (NC & RE) Training Capsule to meet their Renewable Energy needs and support the Armed

Forces' energy saving and regeneration drive (2014–2015). Print and electronic media have highlighted the Force's diligent efforts in operations and disaster assistance in great detail (2015–2015; 2018–2019). An application for the authorisation of LED lighting and Green Building Standards in the Defense Services was started in line with the PM's "Bijli Bachao Desh Banao"[3] policy. Furthermore, Swachhta-based initiatives across the various states are given a great deal of attention each year under the "Swachh Bharat Abhiyan."[4] Additionally, MOD is working on 150 MW Solar Energy Power projects to lessen the country's carbon footprint (Ministry of Defence, 2020). After careful thought, the Government of India decided to enlist the Army to help with the massive task of revitalising the Ganga River as part of the "Namami Gange"[5] project (Defence, 2018, 2021, 2022).

Securitisation of climate change, however, calls for more than just recognising climate change as a concern; it also calls for taking extreme measures to safeguard a referent object from the harm. When the "Annual Reports from Ministry of External Affairs" from 2014 to 2023 were examined more closely, it became clear that the reports met the criteria of the "speech act"[6] concept since they identified climate change as a

---

[3] Dec 14, 2014, an initiative of the Government of India to spread awareness regarding conserving electricity among citizens of India with an out-of-the-box video campaign: "Bijli Bachao, Desh Banao."

[4] Swachh Bharat Abhiyan following initiatives have been taken in this regard: (a) "Swachta Hi sewa" Abhiyan: Citizens Involvement: "Swachhta Hi Seva" campaign was organised by the Units and Institutions of IHQ of MoD (Army); (b) Swachhata Pakhwada: Dec 1–15, 2019 was declared as Swachhata Pakhwada by the Government of India; and (c) Sewage Treatment Plant Projects: To ensure cleaner environment as part of Swachh Bharat Abhiyan, endeavour is being made to provide Sewage Treatment Plants in all stations.

[5] The Ganga Task Force, also known as 137 CETF Bn (TA) 39 GR, has effectively completed projects including biodiversity monitoring, river pollution monitoring, afforestation, riverside construction, and public awareness campaigns. During that time, the battalion started restoration efforts for "Macferson Lake," which had formerly been Prayagraj's largest body of water but had eventually dried up and degraded.

[6] An advocate of this approach, Wæver (Buzan et al., 1998: 26) saw security as a "speech act"—an expression that signifies and acknowledges phenomena as "security," endowing it with particular significance and justifying drastic actions (Buzan et al., 1998: 26). According to this interpretation, the act of uttering is the deed itself (Buzan et al., 1998: 26). As a result, the discourse surrounding security—which recognises a threat and demands immediate action—contains the definition of security. By asserting security status for several concerns and referential objects in the political, military, economic, and

security threat and provided the groundwork for a partial securitisation. The MEA reports emphasise the need for protecting the community's infrastructure, people, property, and society while implying a state-centric approach to security in the goal of creating a resilient community against natural catastrophes (Ministry of External Affairs, 2015, 2016). The recent presidentship of G20, gave India a golden opportunity to showcase its priority and commitment towards climate change. Among the important topics discussed were green development, LiFE (Lifestyle for Environment),[7] inclusive and resilient growth, and the progress made towards the Sustainable Development Goals (Ministry of External Affairs, 2018, 2019, 2020a, b).

If the security agenda has not deemed climate change to be a security concern, a state would not have taken extraordinary measures, which are intentional measures performed beyond political objectives for the welfare of the referent object. A free trade agreement for environmental goods, investment in clean energy projects, reduction of methane emissions, the Climate Change Conference of the Parties (COP), the First India-Japan Environmental Policy Dialogue (2021–2022), the International Solar Alliance (ISA), the Green Strategic Partnerships (2021–2022), BRICS, BIMSTEC, and the SCO Summit are some of the suggested measures to address climate change in the report (Ministry of External Affairs, 2021, 2022). Nevertheless, there is insufficient data to classify the measures as extraordinary. In order to combat climate change, the Prime Minister of India presented the "Five-Point Agenda." The MEA reports had a significantly higher percentage of paragraphs highlighting environmental concerns than the MOD Annual reports. According to McDonald (2012), the securitisation framework places a strong emphasis

environmental domains in addition to the social, environmental, and cultural ones, this social constructivist interpretation of security seeks to expand the security paradigm (Buzan et al., 1998: 1).

[7] To protect and preserve the environment, Prime Minister Narendra Modi introduced the concept of "Lifestyle for the Environment" (LiFE) at COP26 in Glasgow on November 1, 2021. He called on the international community of individuals and institutions to drive LiFE as an international mass movement towards "mindful and deliberate utilisation, instead of mindless and destructive consumption." Everyone has an obligation under LiFE to live a life that is in harmony with the Earth and does not destroy it, both individually and collectively. Under LiFE, those who lead such a lifestyle are recognised as Pro Planet People. "Panchamrita," at the historic UN Climate Change Conference in Glasgow (COP26). India made a commitment to attain "net-zero" carbon emissions by 2070.

on language when creating security, hence language analysis plays a major role in this analysis like "address climate change," "tackle climate change," "global challenge like Climate Change," or "mitigate and adapt to climate change," etc. McDonald (2012) claims that securitisation puts the issues above the level of everyday politics and might make extreme measures possible. However, the policy documents from the Indian government provide a grim picture, and it's evident that the studies don't recommend taking any severe measures.

According to the MEA research, addressing the nation's top strategic threats should take precedence, and climate change is seen as a threat multiplier. For this study, the nation of India and its role in offering solutions to global issues like climate change as well as giving disaster relief and humanitarian help globally, becomes the referent object, on whom extraordinary activities are utilised (Ministry of External Affairs, 2017, 2022). Because of its insistence on adhering to international climate agreements and forming alliances with governments, the international community can also be viewed as a referent object. Further the report's emphasis on energy dominance in the service of economic development, also makes the economy as the security discourse's referent object. The MEA report emphasises India's interest through strategic environmental alliances, while the MOD paper aims to address the wide range of conventional and nonconventional security concerns. It is still debatable, nevertheless, if Indian defence plans are sufficient to handle climatic insecurity. Therefore, the Indian security agenda's securitisation of climate change remains unfinished.

Review of China's Policy Documents

Through a variety of white papers, this section examines the key security policy publications in China. Some of those are: China's National Defence in New Era (CND) (2019); Responding to Climate Change: China's Policies and Actions (RCC) (2022); and China's Green Development in the New Era (CGD) (2023). The white papers are all from President Xi Jinping's government. China's National Defence in the New Era is a document released by the Chinese government that explains the country's defensive national defence policy, as well as the goals, and importance of China's efforts to fortify its military and national defence. In order to actively foster the establishment of a socialist eco-civilisation, China has implemented a national strategy for sustainable development as well as a fundamental national policy of resource conservation and environmental protection, both of which were detailed in the white paper China's Green Development in the New Era (Cordesman, 2019). In order to share its experience and strategies with the rest of the world and to document its success in mitigating climate change, the Chinese government released a white paper titled Responding to Climate Change: China's Policies and Actions.

A significant portion of the policy documents has reflected different environmental challenges in the white paper titled RCC 2022. The common view that climate change is a threat multiplier is supported. In this report, the international community serves as the reference object with climate change being addressed through subheadings such as "Staying Firmly Committed to Green Development," "Building a Fair and Rational Global Climate Governance System for Win–Win Results," and "Transforming the Earth into a Beautiful Home." Although responding to climate change is mentioned as a higher priority in state governance, there is no implementation of extreme measures. Human security structures like food and water security may pose challenges to human security (Paul Bellamy, 2020), and China's Green Development in the New Era specifically addresses food and water scarcity. Individual becomes a legitimate referent object of securitisation in this situation. The report further identifies China's commitment to the concept of a global community of shared future, putting into practice the UN 2030 Agenda for Sustainable Development and a sensible approach to the interplay between development and protection. China has made efforts to reduce pollution, cut carbon emissions, expand green development, and pursue economic growth in tandem with the green transition.

Noting a link between climate change and security concerns does not always lead to the securitisation of climate change (Warner & Boas, 2019). Therefore, even while the two white papers on climate change and green development point out the possible risks associated with it, they do not help to securitise the issue (China, 2021). The white papers (RCC & CGD) only recognise security and other consequences of climate change as future, rather than current, dangers (Xinhua, 2023). The CND white paper only cites the necessity of actively supporting international human-itarian assistance and disaster relief (HADR) efforts, UN peacekeeping operations (UNPKOs), and vessel protection operations as necessary steps towards promoting the nation's sustainable development. The "1 + N"[8] policy framework for carbon emissions peaking and carbon neutrality, China's role as a global leader in the movement to build an Eco-civilisation and a green Belt and Road, and China's active participation in energy transition and efficiency cooperation under the frameworks of G20, ASEAN, EAS, BRICS, SCO, and APEC by adopting a "people-centred approach" are just a few of the climate change initiatives that were discussed (Xinhua, 2023). These policy papers do, however, acknowledge that climate change poses a threat, albeit a possible one, and they obviously do not contain the exceptional actions that should be taken in the cause of security.

COMPARATIVE EVALUATION OF POLICY DOCUMENTS

This comparative analysis offers an illustration of the various security strategies that exist within the international political system, all of which share common ideals and concerns. The individual securitising and de-securitising elements between the security discourses in China and India are compared and examined in this review. Firstly, to contribute to the comparative analysis of the overall percentages of each document that discusses environmental concerns: the percentage of texts describing environmental challenges in the Indian MOD annual report was comparatively lower than that of the MEA annual report. There was a higher overall percentage of paragraphs highlighting environmental challenges in China's security strategies white papers. Although climate change was expressly identified as an existential threat, only a partial securitisation of the issue was noted in the Chinese white papers; however, no extraordinary steps were suggested. One could argue that the political philosophy and other variables like foreign political obligations are to blame for this partial securitisation. However, such routine behaviours do not aid in securitisation, as they pose a risk rather than a threat to security.

The linguistic tone of each policy document is also assessed to determine whether or not climate change has been securitised. In their

[8] When addressing the COP15 gathering on biodiversity held in Kunming, China, during 11–15 October 2021, Chinese President Xi Jinping stated that China would "put in place a '1+N' policy framework for carbon peak and carbon neutrality." "1" refers to the long-term approach to combating climate change and "N" refers to solutions to achieve peak carbon emissions by 2030.

descriptions of the threat posed by climate change, Chinese white papers and MEA Annual reports employ both the past and the future tenses. This portrays climate change as a possible threat that could eventually lead to the creation of a number of security issues. The discourse surrounding resilience and adaptation to natural disasters do not reflect the measures taken. Therefore, these papers/reports don't help to securitise climate change. Each security policy document's referent objects—nation-states, international relations, individuals, and the economy—illustrated in this overview aid in understanding the topics covered in each policy document. The efficacy of the securitisation process increases if referent objects are carefully chosen.

The Human Security Perspective on Climate Change

The degree to which China and India address climate change within the context of human security is examined in this section. "Human security is an approach to identify and address widespread and cross-cutting challenges to the survival, livelihood, and dignity of the people," as stated in General Assembly resolution 66/290. It asks for "responses that increase the safety and empowerment of all people and are people-centred, comprehensive, context-specific, and prevention-oriented." Since it moved the emphasis away from territorial security and onto the main subject: Individual, the human security concept was immediately recognised as a significant change from the then-dominant perspective on security when it was first presented in the 1994 Human Development Report. In a special report on human security, UNDP (2022) notes three additional characteristics of human security: universal, multidimensional, and systemic. These characteristics are particularly relevant today, as concerns affecting people's security become part of a new set of interconnected threats on a planet undergoing dangerous changes due to human pressures.

Given the shortcomings of state-centric approaches to security, talking about climate change from the standpoint of human security could improve security for communities and individuals that are at risk. It was noted in the "AR5 Climate Change 2014: Impacts, Adaptation, and Vulnerability" study that as climate change occurs, human security would increasingly be at risk (Intergovernmental Panel on Climate Change, 2014). Human insecurity virtually never results from a single cause; rather, it is the result of several forces interacting. "Climate change poses a significant threat to human security due to several factors, including (1) the erosion of livelihoods, (2) the compromise of culture and identity, (3) the increase in migration, and (4) the challenges faced by states in establishing the necessary conditions to ensure human security." (Adger, 2014).

Recent events like Tsunami, cyclones, drought, floods make it abundantly evident that livelihood today is vulnerable to abrupt changes in the environment (Bidwai, 2010). It is high time for these kinds of threats to be dealt in human security strategy, with the goal of protecting livelihoods and guaranteeing environmental sustainability. Like an enraged typhoon, a threat to one aspect of human security is likely to spread to all aspects of human security (United Nations Development Programme, 1994), and hence securitisation approach centred on human security covering all of the components need to be taken. The examination of policy reports from China and India indicates that a significant amount of discussion on the security agenda's human security components has been included. By bolstering the national economy, expanding globally, and creating a robust community to withstand natural disasters, Indian Annual reports focus on human security. However, understanding the implications of climate change is not covered.

The Chinese government's white papers on climate change reflect the country's concerns on the inclusion of values from many communities, equality promotion, human rights protection, and individual economic prosperity. This is one of the important approaches to Human security that attaches importance to cultural values. The AR5 Synthesis Report comes to the conclusion that a number of elements in the economy and society, as well as the climate, are strongly related to material components of life and livelihood including food, water, and shelter. India's security policy fails to specifically name individual as referent objects.

In response to the threat posed by climate change, the AR5 Synthesis Report recommends that adaptation techniques be put into practice for human security. Some of these are: diversification of income-generating activities in agricultural and fishing industries, migration as a risk management strategy, and education of women are among the tactics suggested for improving people's well-being. Cultural factors also influence a society's perception of risk, resilience, and adaptability. The MOD reports acknowledge the significance of protecting the environment from degradation, even though they do not emphasise the implications of climate change as done by MEA Annual report. China's white papers attempted to implement indigenously developed Chinese solutions to address climate change.

It is appropriate to use efficient resource management as a strategy to adapt to climate change. To meet the energy needs, the MOD's reports emphasised the promotion of "Non-Conventional & Renewable Energy." To further address climate concerns, the MEA reports and the Chinese Green Policy paper placed a strong focus on forming strategic alliances and working together with other nations. It is however debatable whether, international collaboration will lessen human insecurity or if helping developing nations may infringe on the sovereignty of rich nations. Chinese white policy papers offer a solution-focussed approach to the issue by putting defence systems in place in case the state becomes vulnerable due to climate change. Therefore, even if certain human security implications are acknowledged, China's and India's security policies have not fully addressed human security components necessary to achieve "human" securitisation of climate change.

Conclusion

The analysis concludes that effective extraordinary actions to counter the threat of climate change are absent from the existing security discourse articulated by China and India. To adapt to and lessen the threat posed by climate change, nations must take more proactive efforts. It is necessary for the state to treat climate change as "existential threat" in order to initiate policy actions to address climatic insecurities, since the political community in each country does not explicitly recognise climate change as a pressing security issue. There is more to climate security than just averting natural calamities. The disregard of how climate change may affect human security, could ultimately result in state instability. To some extent, the environmental, adaptability, economic, and political security of the state and the individual have been covered in Indian MEA and MOD reports. However, this approach needs to be expanded beyond acts of resilience building to include environmental challenges, community economic development, and individual political dignity. The Chinese white paper RCC and CGD identified that climate change puts strain on food and water, increases health risks, and affects the economy in detail. However, policy paper CND did not cover any implications of climate change.

"Climate change will lead to new challenges to states and will increasingly shape both conditions of security and national security policies," the AR5 Synthesis Report stated. Many of these climate-related issues will be manageable by robust institutions but without endangering the security of people. Communities, organisations, and civil society must be involved in and supportive of the cause of climate change adaptation and mitigation in order to make progress. Nations may also employ the army to carry out climate change adaptation and mitigation initiatives. Climate change does have an impact on military planning; hence it is strongly advised to increase the military's traditional role without "militarising" a human security. The Indian Army's Namami Gange Project, for instance, is one example of how the military could help safeguard natural resources and habitats.

Therefore, it is imperative to address climate change as a security threat to persons and communities, as it poses an existential risk. In the end, a state-centred security response to climate change is not the only way that climate security should advance. In order to protect our planet Earth for the present and next generations, we must act urgently.

References

Adger, W. J. (2014). Human security. In C.B. Field (Ed.), *Climate change 2014: Impacts, adaptation, and vulnerability. part A: Global and sectoral aspects. contribution of working groups II to the fifth assessment report of the intergovernmental panel on climate change* (pp. 755–791). Cambridge University Press.

Bellamy, P. (2020). Review of the book environmental conflicts, migration and governance. In T. Krieger, D. Panke, & M. Pregernig (Eds.), *Journal of Human Security, 16*(1), 51–52. https://doi.org/10.12924/johs2020.160 10051

Bidwai, P. (2010). *An India that can say yes: A climate-responsible development agenda for Copenhagen and beyond.* Heinrich Boll Foundation.

Buzan, B. (1983). *People, states and fear: The national security problem in international relations.* Wheatsheaf Books.

Buzan, B. (1984). Peace, power and security: Contending concepts in the study. *Journal of Peace Research, 21*(2), 109–125.

Buzan, B. (1998). *Security, the State, the "New World Order." and Beyond. New York:* Columbia University.

Buzan, B., Waever, O., & de Wilde, J. (1998). *Security: A new framework for analysis.* Lynne Rienner Publishers.

China, T. S. (2021, October). *Responding to climate change: China's policies and actions*. Ministry of Ecology and Environment.

Cordesman, A. H. (2019). *China's new 2019 defense white paper: An open strategic challenge to the United States, but one which does not have to lead to conflict*. Retrieved from Center for Strategic and International Studies (CSIS). http://www.jstor.org/stable/resrep22570

Floyd, R. (2007). Human security and the Copenhagen school's securitization approach: Conceptualizing human security as a securitizing move. *Human Security Journal, 5*, 38–49.

Gordeeva, E. (2022, May 10). *The securitization of global environmental policy: An argument against*. Retrieved from European Journal for Security Research. https://doi.org/10.1007/s41125-022-00083-x

Intergovernmental Panel on Climate Change (IPCC). (2014). Human security. In *Climate change 2014: Synthesis report. contribution of working groups I, II, and III to the fifth assessment report of the intergovernmental panel on climate change*. Retrieved from https://www.ipcc.ch/report/ar5/wg2/human-security/

McDonald, M. (2008). Securitization and the Construction of security. *European Journal of International Relations, 14*(4), 563–587. Retrieved from Annual Reports 2018–2019. https://www.mea.gov.in/Uploads/Public ationDocs/31719_MEA_AR18_19.pdf

Ministry of External Affairs. (2015). *Retrieved from annual reports 2014–2015*. https://www.mea.gov.in/Uploads/PublicationDocs/25009_External_Affairs_2014-2015English_pdf

Ministry of External Affairs. (2016). *Retrieved from annual reports 2015–2016*. https://www.mea.gov.in/Uploads/PublicationDocs/26525_26525_External_Affairs_English_AR_2015–16_Final_compressed.pdf

Ministry of Extenal Affairs. (2017). *Retrieved from annual report 2016–2017*. https://www.mea.gov.in/Uploads/PublicationDocs/29521_MEA_ANN UAL_REPORT_2016_17_new.pdf

Ministry of External Affairs. (2018). *Retrieved from annual reports 2017–2018*. https://www.mea.gov.in/Uploads/PublicationDocs/29788_MEA-AR-2017-18-03-02-2018.pdf

Ministry of External Affairs. (2019). *Retrieved from annual reports 2019–20*. https://www.mea.gov.in/Uploads/PublicationDocs/32489_AR_Spread_2020_new.pdf

Ministry of External Affairs. (2020a). *Retrieved from annual reports 2020*. https://www.mea.gov.in/Uploads/PublicationDocs/33569_MEA_annual_Report.pdf

Ministry of External Affairs. (2020b). *Annual report 2020. Government of India*. https://www.mea.gov.in/Uploads/PublicationDocs/32489_AR_Spread_2020_new.pdf.

Ministry of External Affairs. (2021). *Retrieved from annual reports 2021.* https://www.mea.gov.in/Uploads/PublicationDocs/34894_MEA_Annual_Report_English.pdf

Ministry of External Affairs. (2022). *Annual reports 2022.* Retrieved from https://www.mea.gov.in/Uploads/PublicationDocs/36286_MEA_Annual_Report_2022_English_web.pdf

Ministry of Defence. (2015). *Annual report 2014–2015.* Retrieved from Ministry of Defence, GOI. https://mod.gov.in/dod/sites/default/files/AR1415.pdf

Ministry of Defence. (2016). *Annual report 2015–2016.* Retrieved from Ministry of Defence, GOI. https://mod.gov.in/dod/sites/default/files/Annual2016.pdf

Ministry of Defence. (2017). *Annual report 2016–2017.* Retrieved from Ministry of Defence, GOI. https://www.mod.gov.in/dod/annual-report-year-2016-2017

Ministry of Defence. (2018). *Annual reports 2017–2018.* Retrieved from Ministry of Defence, GOI. https://www.mod.gov.in/dod/annual-report-year-2017-2018

Ministry of Defence. (2019). *Annual report 2018–2019.* Retrieved from Ministry of Defence, Government of India. https://mod.gov.in/dod/sites/default/files/MoDAR2018.pdf

Ministry of Defence. (2020). *Annual report 2019–2020.* Retrieved from Ministry of Defence, GOI. https://www.mod.gov.in/dod/node/92433

Ministry of Defence. (2021). *Annual report 2020–2021.* Retrieved from Ministry of defence, GOI. https://www.mod.gov.in/sites/default/files/2020-2021_Annual%20Report.pdf

Minisry of Defence. (2022). *Annual reports 2021–2022.* Retrieved from Ministry of Defence, GOI. https://www.mod.gov.in/sites/default/files/AR_0.pdf

United Nations Development Programme. (1994). *Human development report 1994: New dimensions of human security.* Oxford University Press. https://hdr.undp.org/sites/default/files/reports/255/hdr_1994_en_complete_nostats.pdf

Warner, J., & Boas, I. (2019). Securitization of climate change: How invoking global dangers for instrumental ends can backfire. *Environment and Planning C: Politics and Space, 37*(8), 1471–1488. https://doi.org/10.1177/2399654419834018

Xinhua. (2019, July 24). *China's national defense in the new era.* The State Council, The People's Republic of China.

Xinhua. (2023, January). *China's green development in the new era.* The State Council Information Office of the People's Republic of China.

CHAPTER 10

Tagged: The Role of Social Media in Influencing Environmental Governance in China and India

Divisha Srivastava

Abstract In the twenty-first century, states find themselves facing a unique conundrum: how to communicate with their populace, quickly and effectively. A common solution that has emerged for this is the use of social media to disseminate information. On the flip side, social media has become an integral tool in the hands of the common masses to hold their states accountable. By partaking in these platforms, the state gains access and ease of communication to its subjects, the same tools that are also utilised by those asking questions to power. This paper aims to look at the role that social media has played in advancing environmental security in China and India. The aim for this paper is to find out the differences and similarities in how the two states have had to deal with disseminating information to the public versus dealing with public-driven awareness and mobilisation for environmental causes.

D. Srivastava (✉)
South Asian University (SAU), New Delhi, India
e-mail: dvshasri@gmail.com

Keywords China · India · Environmental Security · Social Media · Mass Communication

INTRODUCTION:

In the twenty-first century one of the major issues facing us today is the health of our environment. States across the globe have been trying to innovate and create new mechanisms through which they can ensure their developmental goals are met while keeping true to the climate change control commitments they've made to their people and to the international system. China and India find themselves in a unique position where their developmental needs are higher than the rest of the world and so are their populations, making this balance challenging to strike. Despite the differences between them, ideologically, politically, economically, and socially, China and India are more similar to each other than they realise. With their rising status as regional power centres to look out for, they both boast a rising urban population, one that is comprised of an informed youth who don't take things at face value. The rise of access to the internet has empowered consumers of information to seek what is not readily advertised, and ask the question why.

In the tech and information age our governments too have been relying on social media platforms to spread and gather information. This holds true especially in developing countries (Chang et al., 2024). Social media platforms have emerged as vital spaces for not only individuals and communities but also for governmental actors and agencies to raise awareness about environmental issues, share information, and mobilise collective action. These platforms provide renewed access to the masses, facilitate the rapid dissemination of news and information about pressing environmental challenges especially those that impact people's livelihoods adversely (Hansen, 2010). Additionally, social media has not only empowered citizens to hold governments and industries accountable for their environmental actions and policies, fostering increased transparency and pressure for more sustainable practices; it has also empowered the government to hold itself, and its many parts accountable.

In China, people have the trust to rely on public social media channels to gather information in case of a disaster. However, the use of social media as a channel of communication while being rampant, isn't uniform across all sectors (Boas et al., 2020). A similar trend is seen in India wherein the government has been using these platforms to spread awareness about its many schemes (Singhal et al., 2018). On the flip side, social media has become an integral tool in the hands of the common man to hold their state accountable. The access that social media accounts and channels provide to the public makes them more accessible for criticisms too.

This chapter aims to look at the role that social media has played in advancing environmental security in China and India. The chapter is divided into four parts, the first part highlights the role of social media in environmental governance, parts two and three showcase China and India's unique experiences with using social media platforms for both the state and civil society when it comes to issues of the environment, and finally part four will compare the experiences of the two states. The aim of this chapter is to find out the differences and similarities in how the two states have had to deal with having a renewed ability for efficiently disseminating information to the public versus being able to deal with the same public being able to use those platforms as an avenue to generate awareness and mobilisation for environmental causes.

The Fifth Estate: Social Media's Resurgence in Matters of Environmental Governance

In 1924 Arthur D. Little espouses the fifth estate as being composed of those who had the 'simplicity to wonder, the ability to question, the power to generalise, the capacity to apply' (Little, 1924, pp. 299). For him and since then the idea of the fifth estate has been one that comprises of thinkers and revolutionaries those that had the courage to question. In 2024, a hundred years later, the same spirit is attributed to the realm of social media and the many stakeholders it platforms. The fifth estate of today however is different in a particular regard than its older forms, today these avenues are used by those in power as well as those who question it.

In contrast to the fourth estate, i.e. traditional mass media of the press and journalism, the fifth estate of social media isn't an end of the fourth nor is it its extension. Rather the fifth estate of new and open media creates a new social reality which manifests itself as a result of the public's capacity to do so (Hidri, 2012, pp. 20). However, as a result of the various geopolitical socio-economic cultural class and racial backgrounds that people come from the public's access to these platforms isn't free and equitable. In spirit the internet is of, by, and for everyone, but in practice that doesn't work out. The digital arena thus emerges as a consequence of games and strategies played by the multiple stakeholders involved which includes both governmental agencies and politicians to lobby groups and activists, all vying to gain access—not just to this public space, but using it to gain access to public's attention (Dutton, 2009, pp. 7).

Nowhere has this tension for public space and attention seen as evidently and acutely as in the realm of environmental governance. According to the World Economic Forum, social media is expertly utilised by the environmental sector in an effort to bridge the gap between what the people know and what states tell them. By using these platforms ordinary people are given access and ability to hold accountable their states and share data with one another. The environment is a shared resource, but it is through social media that individuals find themselves being able to share into the process that governs them by advocating for their shared betterment. The ability to augment collective voices makes it an integral tool for environmental advocacy. Social media thus becomes a mechanism through which multiple stakeholders interact, and share information, which in turn impacts decision-making (Dosemangen, 2016).

These platforms make it easier for both the public and the government to reach one another, unlike in the past; however, in today's world the media consumers hold the power to shift conversations. Individuals, activists, NGOs alike now find themselves in the unique position of being able to garner attention for their specific causes, create communities of support, and organise tangible solutions like arranging financial support to further their cause (Puentes, 2012). Social media today therefore is come to be seen as 'the new "public street," galvanising momentum for change in much the same way as a protest or march, and the more popular the social posts are, the more effective they are in generating action from the government' (EPIC, 2022).

Environmental Governance, Social Media, and the State and Society in China

Unlike its contemporaries particularly the Western countries that compromise largely of democratic institutions which help facilitate the existence of their active civil societies, the People's Republic of China is not a democratic state. However, unlike canonical autocratic states, the Chinese do boast a robust civil society one that is actively engaged in matters of environment. How do the two work together, how does advocacy operate within paradigms that are censored and controlled are integral questions to ask. The internet and thus by extension social media plays an integral role here.

Censorship of social media channels is fairly common in China, this takes the form of two strata: thematic censorship wherein articles are deleted because of the content/issue they are referring to, and persona censorship wherein certain individual actors and activists are blacklisted from posting on social media platforms (Yan & Li, 2023). Despite this authoritarian control, Chinese citizens have continued to find novel ways to bypass the censors and utilise creative methods to voice political criticism without getting flagged. By using different linguistic and technological tactics like puns, emojis, etc. the people find a way to resist and challenge official narratives that are peddled by state agencies (Wu & Fitzgeral, 2020).

By utilising the internet, the government in China has been able to improve the quality of environmental governance by being efficient and on time. Social media has provided citizens with a platform with which they can monitor the government's behaviour and pressurise them to take pro-active action to solve environmental concerns (Wang et al., 2022). Out of the 300 plus million microblogging accounts the Chinese had in 2012, 176,000 accounts were opened and run by the Chinese government and their associated agencies (Zheng, 2013). That number has only increased in the last 10 years and so has their reliance on social media to create a network for information flow. China has a centralised political system which means that all power flows from the centre. However, due to its large size and population despite the CPC's stronghold, it has been difficult to effectively govern. With the help of social media, the central government in China can better supervise local governance. Social media thus becomes a cost-effective conduit for supervision, and people are rewarded with better governance outcomes. Social media exposure from both the centre and the civil society mean that local officials are held doubly accountable for malpractices and if they don't disclose factual information, they suffer the loss of reputation (Gao et al., 2018, pp. 14).

Through social media, citizens gain the potential to voice their concerns, share evidence of environmental harm, and demand accountability from decision-makers. This has led to increased transparency and pressure for more sustainable practices, as governments and industries are now under constant scrutiny from the public. However, unlike the popular adage, all publicity isn't good publicity, especially when it comes to issues of governance in China. What ends up happening instead is that perspectives that are solicited by the state as a part of their public consultations are tolerated during specific time frames (Yan & Li, 2023). For instance, research has shown a positive correlation with efficient governance because both the central and the local governments in China are largely risk-averse. This means that despite being a controlling centralised state, they tend to disclose information, especially during a public crisis under the threat of exposure and loss of reputation (Gao et al., 2018, pp. 16). Both public and private appeals made to government agencies and regulators have shown positive results in reducing environmental concerns. Despite not having a voter-based electoral system to elect those in power, the government in China is held accountable to its people and is expected to deliver to their demands to ensure the legitimacy of their rule (Buntaine, 2022).

In the face of crises, especially, the use of social media has proven to be immensely useful. China has been a digitally forward country for decades now but the 2020 COVID pandemic highlighted the efficacy of social media as a means of communication in the face of a crisis. By using social media channels, the government could disseminate information easily, at a low cost, create an open environment, and foster community participation. What was earlier a singular top-down governance model, with the advent of these platforms, got transformed into a relatively socially driven collaborative governance model wherein the different sectors and stakeholders involved could network with each other and efficiently provide their services or solutions to issues (Zhang & Huang, 2021, pp. 3).

The concept of 'human security' is closely linked with emancipatory ideas of environmental security that call for centring the human in our paradigms in order to provide equitable protections to all people especially because it is often the systemically marginalised communities that suffer the most at the hands of climate crises. While the concept doesn't necessarily directly translate into Chinese rule, there is evidence that suggests that the Chinese concept of security is evolving to become more comprehensive. The 'Chinese Dream' espoused under Xi Jinping aims to care and provide for its 1.3 billion citizens because they believe that all people are 'entitled to live with dignity in a secure environment' (Xiao, 2019). The Chinese government has finessed the ability to utilise selective censorship—they allow (certain) criticisms of the government because they understand that being able to vent and express one's grievances is important for their citizens. It not only helps in creating better governing outcomes; it also solidifies the relationship between the state and its people.

Furthermore, it is also argued that the aim of censorship isn't to quell criticisms of the government (which they often do encourage) but rather to nip potential for collective action in the bud. Therefore, while the government allows for self-expression it doesn't accept the coming together of critical people and ideas. 'The Chinese people are individually free but collectively in chains' (King et al., 2013, pp. 14). Despite what conventional wisdom dictates wherein authoritarian regimes would have absolute control and censorship over social media platforms, in the case of China, the censorship is selective in nature. By allowing some online criticism, the state signals its responsiveness to public demands because the regime and the leader's legitimacy is at stake. In today's day and age where the populace in China is becoming largely urban, educated, and has access to social media, the level of awareness among the masses is high. Fine-tuning censorship to suit regime needs ensures that their citizens can be appeased; it helps in creating popular support and/or helps in stalling large-scale collective dissent (Cairns, 2017). Using their moderators the government has also established for itself a colourful track record of punishing and arresting individuals who use these platforms. The users were labelled to have been 'inciting dissatisfaction with the government' (BBC, 2013). For decades China has been cracking down on grassroots environmental activists despite making environmental protection a state priority (Dyer, 2007).

The citizen's participation however despite social media access for many, remains variable even when the government encourages it. This can largely be attributed to the level of trust that the individuals have over governmental agencies and the efficacy of public mobilisation in garnering results. On one hand thus the government wants people to participate, on the other hand people don't respond as eagerly because of anxiety that their participation can bring. The lack of institutional trust means that despite the availability of platforms and avenues, people are more inclined to not engage with the government (Homburg & Rebecca, 2022). The intertwined relationship between the state and these platforms coupled with its centralised control, punitive measures, and selective censorship add to the inability of people to trust in it. On the other hand, the state could very well just quell social media access in order to maintain a reputable and poised narrative of itself. However, by doing so it would also limit its own ability to be able to learn from the bottom-up—and monitor situations in case they turn threatening. Therefore, the access they allow works in their favour. All and all social media and governance work in tandem only if the central government and citizens share a common goal. If their interests diverge, the local government resorts to policing, and the central government resorts to censorship (Qin et al., 2017, pp. 138).

Environmental Governance, Social Media, and the State and Society in India

India sits at a unique position when it comes to civil society environmental activism because it is the world's largest democracy. Public participation is not only encouraged but is mandated in order to establish the government elected by popular mandate. Being a democracy in the age of social media makes the adage government by the people, of the people, and for the people more potent because the people have the power and access to let their opinions known. Most mass democracies in the world have advanced in tandem with the development of mass communication mechanisms because these are crucial in organising public life and opinion. Furthermore, democracies are the kind of political institutions where both the state and the civil society are seen as active participants. The two find themselves at the opposite ends of the spectrum and thus often find themselves contesting over power (Saed, 2009, pp. 466).

By using social media avenues, individuals have an active way to communicate with governmental agencies, civil society actors have been able to mount successful campaigns to hold the state accountable, create awareness of the environmental impact of upcoming projects, and mobilise masses to oppose cases where the environmental costs cannot be justified. On the other hand, the government and its various agencies find themselves using social media avenues to disseminate information, engage with the masses, and participate in the formation of public opinion by bringing forth state-driven strategic narratives into the common fold.

Activists in India have had a long history of using social media avenues to create awareness regarding impending projects and criticise the government. The environment ministry doesn't have a good track record when it comes to environmental security—activists argue that it has a bias towards promoting developmental projects at the cost of foregoing adequate environmental impact surveys and upholding regulations. The ministry often works in collaboration with corporate entities and doesn't tolerate opposition. Regulatory mechanisms that are put in place to help protect the environment are now coming to be seen as a bureaucratic formality (Bisht, 2019). Furthermore, to add insult to injury the acutely centralised nature of environmental control doesn't bode well for a country as diverse and complex as India (Kashwan & Kodiveri, 2021).

In response to such criticisms that are readily available online, it makes sense that the ministry and agencies present their view points as well. It was reported that the environment ministry often advertises tenders for contracting social media experts to help in precisely this. Agencies are called upon to help the ministry for environment curate their online presence by highlighting their efforts towards meeting sustainable development standards, and 'green' growth over what they are blamed for—destruction in the name of development. Despite having an open line for criticisms, they continue having a presence on these platforms and encourage public participation because these platforms are used as a means to further garner a better sense into public sentiments and maintain channels of communication with their citizens (Aggarwal, 2016). Twelve years ago, the Ministry of Communications and Information Technology released a dossier outlining the guidelines for how different government agencies could utilise social media platforms. The document highlights the merit is using these mechanisms in order to enhance their outreach, be able to have real-time engagement with their target audience, be able to engage with individual users, and finally for managing perceptions especially when it comes to unverified facts and rumours regarding government policies (MoCIT, 2012, p. 7). A key feature of social media platforms is that they simultaneously provide the ability for mass communication and micro targeting. This means that by crafting an efficient grassroots media campaign public opinion both within domestic audiences and for international audiences can be moulded.

Despite having clear guidelines for governmental officials and agencies on how to use these platforms to communicate in an official capacity, 'the democratisation of social media as a political communication platform in India has not meant its professionalisation as an information-sharing one' (Mahapatra & Plagemann, 2019, pp. 2). Political parties and leaders often share information that is not factual and can thus perpetuate false-hoods. Activists and environmentalists often take to these platforms to sound their opposition and fact check wherever necessary. However, they are threatened with the use of anti-terrorist laws against them because their actions are seen as a challenge to 'the sovereignty and integrity of India' (Dutt, 2020). When cases are booked under UAPA, the accused persons are often incarcerated on little to no evidence. The act thus gets used as a form of punishment for the accused without the prosecuting parties having to prove the charges (Venkatasen, 2023). Furthermore, there also exists a trust deficit and a lack of state accountability when it comes to issues of environmental governance because adequate public and civil society consultations are often not undertaken. This means that marginalised groups who fear the loss of their livelihood and are in fear of their survivability are often left in the dark regarding policy decisions, making it all the more challenging for them to trust the state. In order for the state to be able to deliver better policy outcomes, they need to find a way to assuage people's anxieties (Srivastava, 2022, pp. 2).

Enhancing Environmental Governance Through the Use of Social Media

Environmental governance in essence refers to the efficient governance of the environment in order to be able to deal with its diverse issues. It entails entities at different levels from the government and its sister institutions and agencies to NGOs and other community-based organisa-tions. As rising powers in Asia with large populations in the digital age, both China and India find themselves in a precarious position of having to balance the act of information-sharing on platforms that are often actively used to express dissent against them.

As highlighted above, the presence of social media platforms makes it possible for governments to undertake mass messaging at almost zero cost (since these accounts are free to use) and keep their populace well informed, in real time. No other infrastructural communication mechanism can provide this efficiency. This becomes extremely crucial for efficient environmental governance in times of natural emergencies. Human-centred sources of information like microblogging sites like X (formerly known as Twitter) have become a great resource for gathering observational data during emergencies. This helps in crafting better policies to provide relief. Both temporal and spatial issues get solved because real-time information can be shared and geolocation identification can be used to earmark who is in danger and where, facilitating rescue operations (Phengsuwan, 2021, pp. 21). These social media platforms are used to share key textual and photographic information, which in turn can have the potential to improve the conventional information-gathering mechanisms used by states and can thus aid crisis response efforts by providing specificity to earmark affected areas on maps in order to coordinate relief and rescue. These strategies have successfully been used during the Hurricane Katrina in 2005 to the earthquake in Nepal in 2015 (Lazreg, 2018, pp. 8). Other than emergent situations, social media also helps states disseminate information for disaster risk reduction across a broad variety of natural hazards by creating awareness regarding them among the people, sharing what to do in case of emergencies, and guiding people towards right action plans to safeguard themselves (Dufty, 2014).

For society members to the use of social media helps in mitigating disasters. Citizens in both China and India have actively used social media to create community resilience in the face of environmental emergencies. This was made possible because there was a demonstrative ease of convenience in using social media platforms. It enabled members to forge an online community and solidarity with one another, guide each other to navigate unfamiliar situations, and collaborate to solve issues. Access to information helped alleviate anxieties and panic of affected communities (Zhai et al., 2023). For the state community centric and driven source of information help them guide their relief measures in a targeted manner (Bhuvana & Aram, 2019), while for the community state-led information dissemination helps create awareness of the evolving situation (Boas et al., 2020). While accessing personal information can be seen as a breach of privacy, and there are risks and limitations involved in using social media because rhetoric, fake news and cyber-crimes, the usefulness of these platforms for communication purposes doesn't diminish (Giri & Vats, 2019). Thus, it seems imperative not only for governments to allow these platforms for the public but also to maintain a continued and consistent presence of their own as well.

However, having a social media presence and using it for communication comes with its own disadvantages. During the floods in Kashmir in 2014, information regarding relief operations was submerged under harmful rhetoric and propaganda peddled on the same platform for vested political interests (Hooda, 2014). Similarly, during floods in Chennai, 'social media platforms and messaging apps became both conduits and accelerators of unverified information, leading to a cacophony of rumours that compounded the challenges faced by authorities and citizens alike' (Deshkar, 2023). During the COVID pandemic, a similar trend was noticed across the border in China. Sina Weibo (the Chinese version of X/Twitter) was inundated with rumours, conspiracy theories, fake news, and unverified claims. However, unlike the rest of the world, the Chinese social media sites didn't statistically carry narratives that are critical or hold their government responsible. This can be attributed not only to the nature of the state but also because of the immense rumour curbing and myth debunking initiatives undertaken by official agencies and accounts to counterbalance the incorrect information being shared (Song et al., 2021).

When it comes to citizens rights, it is noticed that the 'governments in more liberal and democratic countries, with higher level of media and informational literacy, remain torn between the need to preserve citizens' freedom of speech and curtail the circulation of fake news. However, governments in authoritarian governments are less concerned with the accusation of overstepping the boundaries of free speech (Rodrigues & Xu, 2020). It was reported in 2023 that China shut down over 1 lakh social media accounts in a sweeping effort to quell the spread of misinformation (Orr, 2023).

India on the other hand cannot, in principle, take such drastic steps. Firstly, as a democracy it needs to uphold the freedom of speech for citizens, and secondly the government doesn't have executive control over platforms like X/Twitter. It was reported that the government did implore the company to take down certain accounts. The company saw it as a flagrant violation of the freedom of speech while the government saw it as a violation of the sovereignty of Indian Law that all companies operating in India have to abide by. The government made the company take down accounts because they allegedly peddled fake information and 'anti-India' sentiments; the company stated that a lack of disclosure of information regarding who or what was removed from the platform can lead to lack of accountability and arbitrary decision-making and the opposition called it the 'murder of democracy' (Aljazeera, 2024; Singh, 2023).

CONCLUSION

So where do we go from here? Based on the experiences of China and India in the realm governing the environment through social media channels, a major setback that arises is the lack of trust. There exists a lack of trust between the state and society which exacerbates communication issues especially during environmental challenges. Furthermore, since these platforms are used by those in power to communicate and garner support for political gains, there remains a cynicism regarding state messaging. On the other hand, since these platforms are used by civil society individuals to register dissent, there remains an apprehension to rely on people-driven data sources.

Four parameters can be highlighted to bridge that trust deficit between the state and the society, and facilitate environmental governance—these are responsiveness, transparency, access, and safety. States have to be able to show keen interest to use social media as an avenue to assuage their citizen's grievances. Pro-active participation by the state would encourage citizens to use the platforms to highlight issues where the state has fallen behind. By demonstrating that the state is keen on mitigating these issues, and being open about the progress made or the pitfalls reached, engagement between the state and the society can be held in a transparent fashion. Accountability of one's actions and communication of one's intent would help alleviate stresses and anxieties that the people face. Despite large-scale digitisation and modernisation, many areas don't have access. This is especially pertinent during crises situations because urban populations tend to have better access to the internet than rural or marginalised ones and thus there remains a bias in the data collected (Vieweg, 2014). The state can also provide training and support to citizens, especially from marginalised communities to help them navigate social media platforms and use them effectively for environmental advocacy. By providing better access the state can show its commitment towards safeguarding not only the environment but also citizen's well-being.

This needs to also extend towards citizens rights and safety. On the one hand the state expects citizens to engage and communicate, on the other hand it holds this against them. The state can take measures to protect citizens' personal information and privacy on social media platforms. Equating journalists and environmentalists with terrorists, and punishing them for having dissenting opinions doesn't equate with the stand that China takes in principle as a state that encourages public expression, and India takes as a democracy where freedom of speech is protected constitutionally.

The trust deficit between the state and society poses significant challenges to effective governance, particularly in the realm of environmental management. This trust deficit is evident in both China and India, albeit in different ways due to their distinct political systems. In China, the government's centralised control and selective censorship of social media platforms have led to a situation where citizens are wary of engaging with the state, despite the government's efforts to use social media for governance purposes. In India, the government's attempts to use social media for environmental governance are often met with scepticism and criticism from civil society, which perceives the government as being biased towards promoting developmental projects at the expense of environmental protection.

Without a sense of trust and loyalty, even if given a chance, citizens are less likely to engage with the government, to participate in its decision-making processes, or comply with regulations. This can lead to a situation where the government's policies and actions are not aligned with the needs and preferences of the population, resulting in ineffective governance, subpar mitigation of disaster risks, and a lack of progress on environmental concerns. If used correctly, social media platforms have the potential to play a crucial role in bridging the trust deficit between the state and society. By providing a platform for open and transparent communication, social media can help build trust by allowing citizens to voice their concerns, engage with the government, and hold it accountable for its actions. Additionally, social media can be used to disseminate accurate and up-to-date information on environmental issues, policies, and actions, helping to build public awareness and support for environmental initiatives. As both countries continue to develop and modernise, finding the right balance between open communication and communication control will be crucial for ensuring the effective use of social media in environmental governance.

REFERENCES

Aggarwal, M. (2016, February). Environment ministry's response to criticism social media experts. *The Mint.* https://www.livemint.com/Politics/Dam AXMdwBb0maeqjL92DMP/Indias-environment-ministrys-response-to-critic ism-social.html

Aljazeera. (2024). India's demand to block accounts amid farmers' stir curtails free speech: X. https://www.aljazeera.com/news/2024/2/22/indias-dem and-to-block-accounts-amid-farmers-stir-curtails-free-speech-x

BBC. (2013). *China web users arrested over posts on Sina Weibo.* https://www. bbc.com/news/technology-23795294

Bhuvana, N., & Aram, A. I. (2019, October). Facebook and Whatsapp as disaster management tools during the Chennai (India) floods of 2015. *International Journal of Disaster Risk Reduction, 39.* https://doi.org/10.1016/j. ijdrr.2019.101135

Bisht, D. S. (2019). How the centre is diluting green clearance norms. *Down to Earth.* https://www.downtoearth.org.in/blog/urbanisation/how-the-cen tre-is-diluting-green-clearance-norms-62828#:~:text=The%20poor%20environ mental%20performance%20can,the%20environment%20and%20the%20comm unity

Boas, I., et al. (2020, December). The role of social media-led and governmental information in China's urban disaster risk response: The case of Xiamen. *International Journal of Disaster Risk Reduction, 51,* 101905. https://doi.org/10. 1016/j.ijdrr.2020.101905

Cairns, C. M. (2017). *China's Weibo experiment: Social media (non) censorship and autocratic responsiveness.* Cornell University

Chang, H., et al. (2024). The role of government social media in enhancing environmental governance. *China Economic Journal.* https://doi.org/10.1080/ 17538963.2023.2300865

Dosemangen, S. (2016). Can social media help to save the environment? *World Economic Forum.* https://www.weforum.org/agenda/2016/04/can-social-media-help-to-save-the-environment/

Dufty, N. (2014). *A review of the value of social media in countrywide disaster risk reduction public awareness strategies.* Input paper, United Nations Office for Disaster Risk Reduction.

Dutt, B. (2020, July). How UAPA was used on young environmentalists and media didn't care. *The Quint.* https://www.thequint.com/opinion/how-uapa-was-used-on-young-environmentalists-fridays-for-future-india-and-media-didnt-care

Dutton, W. H. (2009, March). The Fifth Estate Emerging through the network of networks. Promestheus.

Dyer, G. (2007). *Activist held in China pollution battle.* https://www.ft.com/ content/a3e99942-41ea-11dc-8328-0000779fd2ac

EPIC Energy Policy Institute at the University of Chicago. (October 2022). *Social media engagement increases government action, decreases pollution.* https://epic.uchicago.edu/news/social-media-engagement-increases-government-action-decreases-pollution/

Gao, S., et al. (2018). The role of social media in promoting information disclosure on environmental incidents: An evolutionary game theory perspective. *Sustainability, 10*(12), 4372. https://doi.org/10.3390/su10124372

Giri, D., & Vats, A. (2019). Social media and disaster management in India: Scope and limitations. In *Smart technologies and innovation for a sustainable future.* https://doi.org/10.1007/978-3-030-01659-3_41#:~:text=Due%20to%20good%20reach%20of,Facebook%20and%20other%20blogging%20sites

Hidri, A. (2012, November). The fifth estate: Media and ethics. *Journal of Arab and Muslim Media Research, 5*(1).

Homburg, V., & Rebecca, M. (2022). Weibo to the rescue? A study of social media use in citizen–government relations in China. *Transforming Government: People, Process and Policy, 16*(1).

Hooda, D. (2014). *India, J&K floods and the social media: The good, the bad and the ugly.* https://reliefweb.int/report/india/jk-floods-and-social-media-good-bad-and-ugly

Kashwan, P., & Kodiveri, A. (2021). Who will guard the guardians? State accountability in India's environmental governance. *Economic and Political Weekly.* https://www.epw.in/engage/article/who-will-guard-guardians-state-accountability

King, G., et al. (2013, May). How censorship in China allows government criticism but silences collective expression. *American Political Science Review,* 1–18.

Little, A. D. (1924, October). The fifth estate. *Science, 60*(1553), 299–306. https://www.jstor.org/stable/1650438

Mahapatra, S., & Plagemann, J. (2019). *Polarisation and politicisation: The social media strategies of Indian political parties.* German Institute of Global and Area Studies (GIGA). https://www.jstor.org/stable/resrep24806

MoCIT. (2012). Ministry of Communications and Information Technology Framework & Guidelines for Use of Social Media for Government Organisations. https://www.meity.gov.in/writereaddata/files/Social%20Media%20Framework%20and%20Guidelines.pdf

Orr, B. (2023, May). China shuts 1 lakh social media accounts in efforts to stop 'fake news' & rumours. *The Print.* https://theprint.in/world/china-shuts-100000-fake-news-social-media-accounts-ramps-up-content-cleanup/1578435/

Qin, B., et al. (2017, Winter). Why does China allow freer social media? Protests versus surveillance and propaganda. *Journal of Economic Perspectives, 31*(1), 117–140.

Rodrigues, U. M., & Xu, J. (2020). Regulation of COVID-19 fake news infodemic in China and India. *Media International Australia, 177*(1), 125–131. https://doi.org/10.1177/1329878X20948202

Saed, S. (2009, June). Negotiating power: Community media, democracy, and the public sphere. *Development in Practice, 19*(4/5), 466–478. https://www.jstor.org/stable/27752087

Singh, M. (2023, June). India threatened to shut down Twitter and raid employees' homes, Jack Dorsey says. https://techcrunch.com/2023/06/12/india-twitter-jack-dorsey/#:~:text=During%20the%20protests%20in%20January,situation%20based%20on%20fake%20news

Singhal, P., et al. (2018). Modern social media in environmental management and sustainability. *Handbook of Environmental Materials Management.*

Song, Y., et al. (2021, December). The "Parallel Pandemic" in the Context of China: The spread of rumors and rumor-corrections during COVID-19 in Chinese social media. *American Behavioral Science, 65*(14), 2014–2036. https://doi.org/10.1177/00027642211003153

Venkatasen, V. (2023, October). Why UAPA is a threat to media freedom in India. *The Hindu.* https://frontline.thehindu.com/columns/uapa-media-freedom-newsclick-china-terrorism-modi-government/article67388663.ece

Vieweg, S., et al. (2014). Integrating social media communications into the rapid assessment of sudden onset disasters. https://doi.org/10.1007/978-3-319-13734-6_32

Wang, B., et al. (2022, December). Government responsive selectivity and public limited mediation role in air pollution governance: Evidence from large scale text data content mining. *Resources, Conservation and Recycling, 187*, 106553 https://doi.org/10.1016/j.resconrec.2022.106553

Wu, X., & Fitzgeral, R. (2020, May). 'Hidden in plain sight': Expressing political criticism on Chinese social media. *Discourse Studies 23*(3). https://doi.org/10.1177/1461445620916365

Xiao, R. (2019). Human security in practice: The Chinese Experience. *Human Security Norms in East Asia*, 45–65, https://doi.org/10.1007/978-3-319-97247-3_3

Yan, X., & Li, L. (2023). *Censoring the intellectual public space in China: What topics are not allowed and who gets blacklisted.* Cambridge University Press. https://www.cambridge.org/core/journals/perspectives-on-politics/article/censoring-the-intellectual-public-space-in-china-what-topics-are-not-allowed-and-who-gets-blacklisted/B5774AC7925D68814C989326EC3AE36B

Zhai, L., et al. (2023). Harnessing ICT resources to enhance community disaster resilience: A case study of employing social media to Zhengzhou 7.20 Rainstorm, China. *Water, 15*(19), 3516. https://doi.org/10.3390/w15193516

Zhang, X., & Huang, R. (2021). *The role of social media in public crisis governance.* E3S Web of Conferences, 253. https://doi.org/10.1051/e3sconf/202125301066

From Policy to Practice: China's NEV Initiatives and the Evolution of Ecological Civilisation

Akhilesh Kumar and *Varaprasad S. Dolla*

Abstract China's NEV industry, a global leader, is fuelled by policies aligning development with ecological consciousness. This chapter examines the interplay between China's NEV policies and the ecological civilisation philosophy (生态文明 Shengtai wenming), focussing on greening policies, sustainability, and implementation priorities. Ecological civilisation, as emphasised in President Hu Jintao's 2007 report, integrates green development principles, including financial sector guidelines and public participation. The NEV industry, a key component of this green trajectory, aims for a 12kWh/100km fuel economy by 2025 and 20% of new vehicle sales as NEVs. This study explores how state-led NEV initiatives transform green mobility, blending economic growth and environmental stewardship, and shaping China's narrative of sustainability versus development. This paper further hypothesises that

A. Kumar (✉) · V. S. Dolla
Centre for East Asian Studies, School of International Studies, Jawaharlal Nehru University, New Delhi 110067, India
e-mail: akhile42_isb@jnu.ac.in

© The Author(s), under exclusive license to Springer Nature Singapore Pte Ltd. 2024
J. T. Karackattu et al. (eds.), *Environmental Securitisation in India and China*, https://doi.org/10.1007/978-981-97-9160-6_11

179

China's NEV industry development policy exemplifies blend of economic growth and environmental stewardship, embodying the principles of ecological civilisation. China is positioned to lead the shift to a more environmentally friendly automotive future while furthering its vision of ecological consciousness and green development by adopting sustainable practices, encouraging innovation, and placing a high priority on public engagement..

Keywords Ecological Civilisation · Green Development · New Energy Vehicles · Sustainability · Ecological Consciousness

Introduction

In the midst of global efforts to address climate change, foster and strengthen sustainable development, China has emerged as a leading force in advancing green mobility through its New Energy Vehicles (NEVs) industry. Rooted in comprehensive policies that intertwine development with ecological consciousness and green development, China's NEV initiatives represent a pivotal component of its evolving approach to environmental stewardship. This paper delves into the intricate nexus between China's NEV industry development policy and its overarching philosophical framework of ecological civilisation (Shengtai wenming 生态文明), illuminating themes of greening policies, the nuanced interplay between development and sustainability, and the prioritisation of ecological civilisation's implementation. At the core of China's commitment to ecological civilisation lies President Hu Jintao's seminal 2007 report to the 17th National Party Congress, which serves as a beacon for transformative change. This ideology seamlessly integrates various elements essential for green development, exemplified by the adoption of common green guidelines in the financial sector under the ecological civilisation umbrella and the emphasis on public participation in shaping China's ecological trajectory. Central to this narrative is the NEV industry, showcased by its exponential growth and ambitious targets outlined in the New Energy Automobile Industry Development Plan (2021–2035). With objectives ranging from achieving specific fuel economy benchmarks for electric vehicles to increasing the market share of NEVs, this plan underscores not only technological innovation and structural adjustment but also sustainable resource management, ecological protection, and the fortification of regulatory systems. Drawing on policy documents related to NEVs, this paper scrutinises how state-led initiatives are reshaping the socio-economic landscape of green mobility, paving the way for a 'green' path of development in China. Furthermore, it explores the dynamic interaction between the official narrative on ecological civilisation and the core tenets of green development, particularly in navigating the discourse of 'sustainability' versus 'development'. Ultimately, this paper hypothesises that China's NEV industry development policy represents a harmonious blend of economic growth and environmental stewardship, embodying the fundamental principles of ecological civilisation. Through a concerted effort to embrace sustainable practices, foster innovation, and prioritise public engagement, China is poised to lead the transition towards a greener automotive future while advancing its vision of ecological consciousness and green development.

Based on the introduction of ecological civilisation and NEVs Policies, this paper hypothesises that; China's NEV industry development policy embodies a blend of economic growth and environmental stewardship, and the policy exemplifies the principles of ecological civilisation through sustainable practices, innovation, and public engagement, and China's position to lead the transition towards a greener automotive future.

Conceptualising Ecological Civilisation (Shengtai Wenming 生态文明)

After the crucial phase of reform and opening up, environment and ecology in China was viewed as a critical source of development and the discourse on ecological civilisation was seen as a Marxist response of the unhindered capitalist development which solely focusses on the material development on the price of environment and ecology. In their several policy documents, officials termed the concept of ecological civilisation as a "harmonious relationship between human and nature" (Zhou, 2020, page 85). Apart from theoretical level, on the policy level, China held the first conference on environmental protection, termed as the 'National Working Conference on the Environmental Protection 2018' in the Beijing where President Xi Jinping emphasised that the Communist Party of China (CPC) should lead the construction of ecological civilisation (Diagram 11.1).

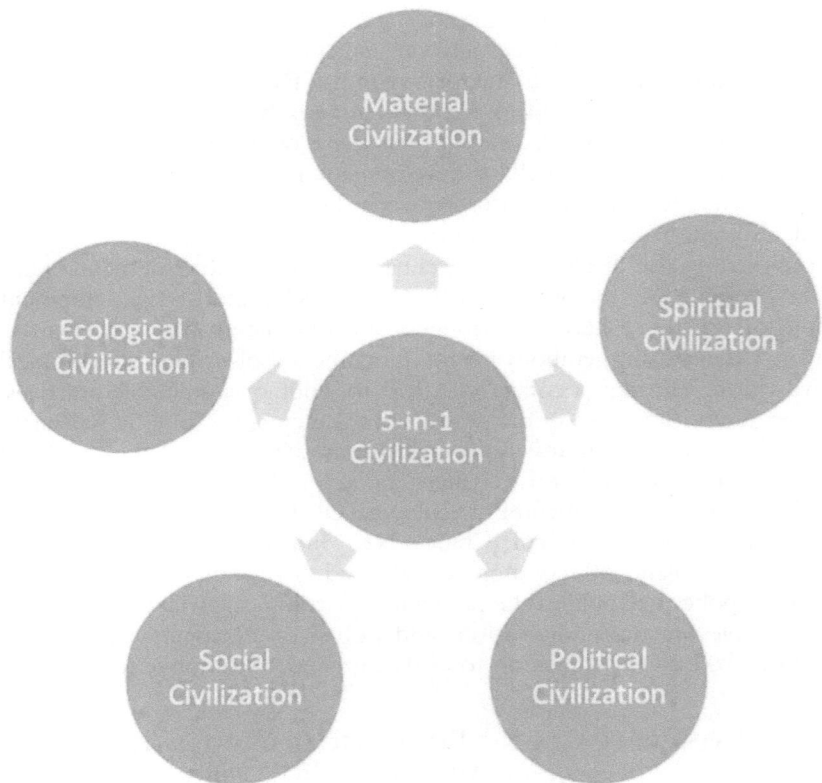

Diagram 11.1   Five-in-one civilisation in China (Authors' illustration)

During the conference, Xi Jinping emphasised the integral connection between ecology, environment, and people's well-being, urging concerted efforts to meet public expectations by bolstering the development of ecological civilisation and delivering high-quality ecological products. In the aforementioned conference, President Xi outlined key principles to advance ecological civilisation in the new era (Liangyu, 2018):

1. Harmony between Human and Nature: Upholding the principle of prioritising conservation and protection, China must rely primarily

on natural recovery to achieve harmony between human activities and the environment.

2. Value of Natural Assets: Recognising the invaluable nature of clear waters and lush mountains, China should strive for innovative, coordinated, green, and inclusive development, fostering resource-saving and environmentally friendly spatial patterns, industrial structures, and living modes to facilitate natural ecological rehabilitation.

3. Well-being Through a Sound Ecological Environment: Addressing pressing environmental issues detrimental to public health is paramount, aligning with the populace's increasing demand for a healthier environment.

4. Integrated Approach to Ecological Preservation: Viewing Mountains, rivers, forests, farmlands, lakes, and grasslands as interconnected life communities, China must adopt comprehensive planning and multifaceted measures in building ecological civilisation.

5. Institutional Arrangements and Rule of Law: Strengthening institutional innovation and law enforcement is essential for effective environmental protection, necessitating the strictest adherence to regulations.

6. Global Collaboration for Ecological Civilisation: China commits to actively participating in global environmental governance, seeking international cooperation to address climate change and contribute to worldwide solutions for environmental protection and sustainable development.

Xi Jinping's outlined principles highlight the importance of integrating ecological considerations into national development strategies, ensuring a sustainable and harmonious relationship between human activities and the environment. By adhering to these principles and implementing robust institutional mechanisms, China aims to make significant strides in advancing ecological civilisation both domestically and globally.

"Ecological Civilisation" is a vital part of realising the Chinese Dream of President Xi Jinping (Xi, 2017). After all, economic development is intertwined with ecological and environmental sustainability. Environmental and ecological health is a matter of public concern in Post-Reform China. Due to its degraded ecological and environmental situation, Pan Yue, the then Vice Minister of China's State Environmental Protection Administration (SEPA), cautioned in 2005 that the economic miracle would end soon because the environment could no longer keep pace (Economy, 2007).

RATIONALE BEHIND THE PHILOSOPHY OF "ECOLOGICAL CIVILISATION"

After ushering in economic reforms and opening up (改革开放 Gǎigé kāifàng) in 1978, China's gross national product (GNP) achieved a sustained annual growth of around 10 per cent. According to the International Monetary Fund's (2020) World Economic Outlook report presented in October 2020, China has now displaced the United States (USA) to become the largest economy in the world when measured by purchasing power parity (PPP). For China, however, this rapid economic growth has come at a high cost and was accompanied by deterioration and degradation of its environment. The high-growth, resource-intensive development strategy China has pursued, coupled with the norms and institutional relationships designed to support this development strategy, has played a critical role in the deteriorating quality of the environment (Jahiel, 1998). Environmental degradation has emerged as one of the Chinese government's pressing challenges. The environmental governance structure in China is characterised by authoritarian environmentalism(Xiang & Lo, 2024) and, more specifically, 'decentralised authoritarianism'. While the Chinese government has recognised the challenges posed by a deteriorating environment and introduced a plethora of rules, regulations and initiatives geared towards environmental protection, the characteristic features of this environmental governance structure ensure that all these efforts have only delivered mixed results. This ecological ethos of governance can be seen in the ideological stream of President Xi; as he put it, "Authorities must work to balance environmental protection and economic growth… to achieve social and economic benefits while maintaining a clean environment" (Xi, 2017).

ECONOMIC COST OF ENVIRONMENTAL DEGRADATION IN CHINA

Environment and ecology are the critical components of the resource distribution of political economy; minimal sustainable policy towards this fraction can harm the whole economy. Eminent Sinologist Elizabeth C.

Economy termed this environmental degradation to its economy as the "Great Leap Backward" (Economy, 2007). She argues that China's environmental problems are worsening, and the country is quickly becoming one of the world's major polluters. The situation is deteriorating because, even when Beijing sets high environmental goals, local leaders typically ignore them, preferring to focus on accelerating economic growth (Economic, 2007). However, she advises that Revolutionary bottom-up political and economic reforms will be required in China to improve the environment truly. Further, she reiterated that the reform of the economic front must be continued on the ecological and environmental front. However, Prof Economy pointed out the problem of economic inefficiency through mismanagement of the environment and ecology in China. Still, a methodology was the unavailability to measure this economic cost. Therefore, in this regard, Guoxia Ma and Others (2020) evaluated the economic cost of environmental degradation in China using the conventional environmental economic methodology from 2004 to 2017. In this research, Guoxia Ma and others (2020) have summarised change in both causes and costs of China's environmental degradation.

The previous research has revealed the following results:

1. From 2004 to 2017, China's environmental degradation cost climbed from 511 billion yuan to 1,892 billion yuan, although its proportion of GDP fell from 3.05 per cent to 2.23 per cent;

   1. The environmental degradation cost growth rate was lower than the GDP growth rate. The increase in environmental degradation costs has dropped dramatically, from 10% in 2014 to 2% in 2017.
   2. The provinces of Shandong, Hebei, Jiangsu, Henan, and Guangdong had the highest environmental degradation costs. In Jiangsu, Guangdong, and Zhejiang, the annual average increase rate of ecological degradation expenses was lower than their respective GDP growth rates.
   3. Considering the cost of environmental degradation in decision-making could help China achieve high-quality development.

ECOLOGICAL-CONSCIOUSNESS (ECO-CONSCIOUSNESS) AS A PREREQUISITE FOR THE ECOLOGICAL CIVILISATION

The term *eco-consciousness* deals with the burgeoning consciousness related to the issues of environment or ecology. This eco-consciousness has been a part of the overall Chinese policymaking process in the post Reform era. However, issues related to the environment are always tackled as nonessential. The environmental Governance structure in China resonated well with President Xi's thoughts on the management of the environment and ecology. President Xi expressed his thought as follows:

> China should ensure the complete, accurate and comprehensive implementation of the new development concept, maintain the strategic resolve and plan economic and social development with a lofty goal of achieving harmonious coexistence between humans and nature. (Xi, 2017)

China suffers from various environmental issues, including air pollution, water scarcity and pollution, land degradation and desertification, and alarming soil pollution levels. According to an Asian Development Bank (2012) study titled 'Toward an Environmentally Sustainable Future: Country Environmental Analysis of the People's Republic of China', less than 1 per cent of China's five hundred largest cities meet the World Health Organization's air quality standards(Asian Development Bank, 2012)... One of the significant causes of air pollution in China continues to be excessive reliance on coal to meet its energy needs. Since 2011, China has consumed more coal than the rest of the world combined, and in 2019, coal made up 57.7 per cent of China's energy use (China Power, 2016). Similarly, a 2014 government survey showed nearly a fifth of the country's farmland was contaminated by chemical waste, pesticides, mining residues, and heavy metals (Stanway, 2019). China's water pollution is as bad as its air pollution, as suggested by a 2017 report published in The Guardian titled 'In China, the water you drink is as dangerous as the air you breathe,' according to which a Greenpeace research has shown that nearly half the country has missed its five-year water quality targets (Deng, 2017). This reflects the severity of the environmental crisis in China. As Elizabeth Economy (2007) puts it succinctly in her article, 'The Great Leap Backward? The Costs of China's Environmental Crisis' written for Foreign Affairs,

'China's environmental problems are mounting. Water pollution and water scarcity are burdening the economy, rising levels of air pollution are endangering the health of millions of Chinese, and much of the country's land is rapidly turning into desert.' China has become a world leader in air and water pollution and land degradation and a top contributor to some of the world's most vexing global environmental problems, such as the illegal timber trade, marine pollution. The Chinese government has recognised both the severity of environmental degradation and the urgent need for efforts geared towards environmental protection, as evidenced by several decades of legislative and policy interventions. Beginning in the 1970s with the establishment of the Environmental Protection Leadership Group, China embarked on a journey marked by significant milestones in environmental governance. Participation in the inaugural United Nations Conference on the Human Environment in 1972 signalled a nascent commitment to international environmental cooperation, followed by the formal declaration of environmental protection as a state policy in 1983.

Since the 1980s, China has enacted a comprehensive array of policies, laws, regulations, and measures aimed at environmental protection. Noteworthy among these initiatives is the formulation of ten significant measures in August 1992 following the United Nations Conference on Environment and Development. This was further complemented by the approval and promulgation of China's Agenda 21 in March 1994, which outlined the nation's overarching strategy and programme of action for sustainable development. The subsequent establishment and evolution of governmental bodies such as the State Environmental Protection Administration (SEPA) and later the Ministry of Environmental Protection underscored China's institutional commitment to environmental governance.

China's legislative landscape is characterised by a plethora of environmental laws addressing various facets of pollution and resource management, including laws targeting water and air pollution, solid waste management, marine environment protection, and wildlife conservation, among others. The revision of the Environmental Protection Law in 2014 expanded the scope of environmental regulation, reflecting ongoing efforts to adapt to evolving environmental challenges. Despite the abundance of legislative and policy measures, the efficacy of China's environmental governance framework remains subject to scrutiny. The prevailing authoritarian environmentalism in China, characterised by centralised decision-making and limited public participation, poses challenges to effective environmental governance. While this model may yield extensive policy outputs, as evidenced by the extensive legislative corpus, its ability to translate into tangible environmental outcomes is contested.

Scholarly discourse on China's governance structure underscores the dichotomous nature of its impact on economic growth and environmental protection. Chenggang Xu's (2011) analysis highlights the regionally decentralised authoritarian system, wherein subnational governments wield significant influence over economic activities, shaping economic growth and environmental degradation trajectory. Elizabeth Economy's critique extends to the institutional configuration's inherent emphasis on economic development at the expense of environmental protection, driven by a pervasive ethos of consumerism and profit maximisation.

Beyond China, similar governance dynamics are observed across the Global South, where authoritarian environmentalism intersects with developmental imperatives. Countries such as India and Brazil grapple with analogous challenges, marked by rapid industrialisation and urbanisation, resulting in environmental degradation and public health concerns. Despite legislative and policy interventions, weak enforcement mechanisms and limited public participation hinder progress towards sustainable development.

In other words, the journey towards environmental sustainability in China is characterised by a complex interplay of legislative, institutional, and socio-economic factors. While legislative measures reflect a commitment to environmental protection, their effectiveness is contingent upon addressing implementation gaps and cultivationg and strengthening meaningful public participation in decision-making processes and climate change. As China's pollution woes increase, so too do the risks to its economy, public health, social stability, and international reputation (Economy, 2007).

The NEV Industry in China

Reforms and development in the New Energy Vehicles (NEVs) sector are closely related to energy security and deteriorating environmental conditions. Its rapid development in the past four decades has added substantial environmental costs and other health-related issues. Apart from environmental issues, China is the world's largest crude oil importer (U.S. Energy Information Administration, 2024, April 16), which endangers its energy security and makes China prone to fluctuations in the international crude oil market. Thus, China's optimum choice is the green fuel energy vehicle.

The automobile industry has been seen as a vital driver of economic growth in the last few decades. With its supply chain, R&D, auto parts production and after-maintenance services, the automobile industry has many positive effects on the economy. Also, it supports various upstream and downstream industries such as electronics, chemicals, steel, etc. Thus, the automobile industry creates a spillover effect on the economy (Yvette To, 2024). The development of the automobile industry is essential for any economy as it creates several employment opportunities.

Today, China is the largest EV market (EVBoosters, 2024, September 27); however, on the eve of the market reforms, the Chinese automotive industry was unproductive and a decade behind the global standard. Thus, Deng Xiaoping, the architect of modern China, chose this industry as a 'pillar' industry (Thun, 2006 cited in Yvette To, 2024). In the initial phase, indigenous automakers needed protection from foreign manufacturers, but at the same time, they required capital and investment. Thus, the Chinese government adopted a new strategy for automotive development called 'Market Access for Technology' (To, 2024: 131). Through Market Access for Technology, foreign investment in the automobile sector was allowed into China through partnerships with State-Owned Enterprises (SOEs). China's automobile sector attracted considerable financial and technical investment in this process. China's admission to the World Trade Organization (WTO) in 2001 proved to be a thrust for the auto sector as China's domestic automobile sector had deeply integrated with the global network.

China's shift towards NEVs contains several administrative as well as ideological components. In the 13th Five Year Plan 2016–2020, 'green development' was identified as one of its core developmental ideologies (Hu et al., 2018).

Feng, Suling et al. (2022) explore the link between digital finance and innovation inequality in the context of China's growing income inequality. The authors focus on green technological innovations as a crucial factor for achieving high-quality economic growth and reducing the gap in green technological innovation between regions, which is essential for addressing regional development inequality in China. The study found that digital finance significantly promotes green technological progress in underdeveloped regions, increases green technological innovation capabilities, and reduces market segmentation. Moreover, digital financial development can encourage investments and alleviate inequality in the green credit market.

Murat Peksen (2021) discusses the correlation between addressing climate change and introducing green energy sources, such as NEVs, powered by hydrogen energy. With the significant increase in CO_2 emissions (about 24.1% contributed by China alone), decarbonisation has become crucial for maintaining a clean and healthy environment. Peksen proposes a three-stage strategic process for transitioning to a hydrogen economy through NEV adoption: gradually reducing fossil fuel consumption, initiating hydrogen generation, distribution, and utilisation, and consolidating the transition into everyday life. The Chinese government has also introduced a three-phase policy for transitioning to a hydrogen economy, starting from 2015, which includes demonstration, popularisation and application, and large-scale promotion. Mastering key technologies is essential for combating climate change, and EVs powered by hydrogen are more efficient than ICEs powered by fossil fuels. The author also addresses social, ecological, technological, and economic challenges that may arise during the adoption of NEVs.

Deteriorating environmental condition has been a severe threat to economic development in China. In the background of environmental degradation, Hu et al. (2021) analyse the impact of the technology revolution on NEVs and their relation with air quality in China. In the past, several researchers have shown that the innovation of hybrid and plug-in NEVs can effectively alleviate CO_2 emissions (Tang et al, 2011) (Su et al., 2021) and (Zahoor et al., 2023) .One of the research papers used a 'panel fixed effect model' in which the data from 31 provinces was analysed to show the relationship between technological innovation of NEVs and air quality. The primary result of this paper is that the impact of technological innovation of NEVs on air quality is not statistically negative. However, Hu et al. argue that technological innovation of NEVs has more impact on the air quality in the regions with higher Vehicle and Vessel Tax (VVT). The study suggests that the companies and automotive industry should increase expenditure in R&D (in technological up gradation of NEVs technology), which would be an excellent means to improve air quality.

Industrial policymaking is essential in differentiating between a planned economy and an economy that relies only on the market. In the light of the above statement, in their paper, Liu et al. (2020) analyse whether China's industrial policy is effective. For this empirical study, the statistical data (of NEV policy) from 2006 to 2018 has been analysed. Furthermore, this paper has identified a positive correlation between technological progress and NEVs development. It also posits that the current state of NEVs patents denies the assumption that China is leading in NEV Technology. Further, this paper argues that promoting a 'high-intensity' policy in the NEV sector is necessary to realise technological upgradation.

Su et al. (2021)explore how NEVs can help achieve carbon neutrality targets. However, analysing the impact of NEVs and pollution reductions regarding PM2.5, Su et al. (2021)asserted that NEVs cannot be considered an efficient measure to mitigate air pollution. Nevertheless, cost reduction of NEVs and government subsidies coupled with investments in R&D into NEV technologies can accelerate this process.

As a part of the new and critical technologies for NEVs, Fuel Cells (FC) proved to be more efficient than conventional Internal Combustion Engines (ICEs), and FCs emit zero-carbon emission. In this context, Kendall (2018) argues that FCs have the potential to play a vital role in the transition to a more sustainable transportation system. Further, he posits that FC technologies like PEMFC and SOFC (Proton-exchange Membrane Fuel Cell and Solid Oxide Fuel Cell, respectively) offer more air quality advantages over traditional propulsion technologies. Thus, China has a significant potential to become a major producer in the area of FC technologies; however, scaling up FC technologies and commercialising them into the global market are arduous.

NEVs Policies in China

For the automobile sector, there are two specific policies which have been introduced by the state council and respective state agencies. This paper will evaluate two policies related with New Energy Vehicles industry. First plan of our study is the *Energy-Saving and New Energy Vehicle Industry Plan for 2012 to 2020* and the second one is *New Energy Vehicle Industrial Development Plan for 2021 to 2035*.[1] Both these plans were introduced by the State Council. For the Plan 2021–2035, state council aimed to form a globally competitive auto industry with advanced technologies related with New Energy Vehicles. Also transition to an energy-efficient

---

[1] State Council, "Notice on printing and issuing the development plan for the new energy vehicle industry (2021–2035)" [国务院办公厅关于印发新能源汽车产业发展规划(2021–2035年)的通知] (2020), http://www.gov.cn/ zhengce/content/2020–11/02/content_5556716.htm.

and carbon society was characterised in the Plan of 2021–2035. Further it is pledged that the improvisation of the national energy security and of the air quality, climate situation would be mitigated and this would lead to the economic growth further. There were some sharp contrast difference and similarities also available in the Plan 2012–2020 and Plan 2021–2035. As the Plan 2012–2020 was related with the NEVs as well as ICE (Internal Combustion Engine) vehicles, however, Plan 2021–2035 is solely related with the emerging New Energy Vehicles technology. This clearly shows the importance and criticality of green mobility in the changing geo-political landscape.

Four Modernisation Theory and Ideological Syncretism Towards NEV Industry

China's Four Modernization theory (四个现代化), proposed by Premier Zhou Enlai (1954-1976) in 1963, dictates a comprehensive roadmap for socio-economic development across four key sectors including agriculture, industry, defense, and science and technology (Pruden, 1986). Over time, this framework has evolved to address contemporary challenges and opportunities, as evidenced by Li et al.'s (2022) conceptualisation of the New Four Modernizations, which emphasises new urbanisation, industrialisation, informatisation, and agricultural modernisation. Integrating ecological concerns and technological advancements has become increasingly crucial within this evolving paradigm, particularly in light of China's commitment to building an ecological civilisation.

In light of the above facts, the NEV industry stands at the intersection of several facets of modernisation outlined within China's developmental framework. This paper delves into the intersection of socio-economic development and NEV policies in line with the Four Modernization Theory. Subthemes of 'Technological Modernisation and Innovation' resemble the objective of this paper.

Evidently, from various policy documents, technological innovation is a linchpin in the New Energy Policy in China. The shift towards electric vehicles (EVs) represents a paradigmatic advancement in automotive engineering, requiring substantial investments in research and development. This aligns with the Four Modernizations' emphasis on science and technology as a cornerstone of national development. The NEV sector fosters indigenous technological capabilities and promotes cross-sectoral collaboration, driving advancements in battery technology, charging infrastructure, and smart mobility solutions.

The widespread deployment of NEVs signals the beginning of an important transformation in China's industrial structure, which will see a move towards greener and more sustainable production methods. Traditional car makers have to reconsider their production methods and product lines as the government provides legislative measures and subsidies to encourage the adoption of NEVs. This strategic realignment, however, within the framework of a greener and more resource-efficient economy reflects the Four Modernizations' mandate of industrial modernisation. Also, with the deployment and development of NEVs, environmental conservation would be achieved. By adopting NEVs, China is moving towards a more sustainable transportation ecosystem. Also, it resonates with the ecological civilisation principles espoused by President Xi Jinping.

The development and deployment of NEVs also profoundly impact the urban landscape. As urban China has been grappling with congestion, pollution and energy security concerns, adopting electric vehicles will ease this distress. The convergence of China's Four Modernizations theory and the expanding NEV industry underscores the country's proactive stance towards embracing innovation, sustainability, and inclusive development. Integrating the NEV sector within the larger modernisation framework presents great opportunities for advancing China's socio-economic goals and paving the way for a more just and environmentally harmonious society as it develops and transforms. Here is a brief table about how the Four Modernization Theory characterises development in the NEV sector: (see Table 11.1)

Sustainable Development Goals and the NEV Policies: How Is the NEV Industry Helping Achieve These Goals

China's commitment to the Sustainable Development Goals (SDGs) of the United Nations is a testament to its dedication to sustainable development. China is a growing economic superpower and the world's largest emitter of greenhouse gases. As such, its pursuit of sustainable development has significant consequences for global environmental conservation

Table 11.1   Four Modernization Theories & Their linkages with NEV Policies

Aspects of Four Modernisation	Linkages with NEV Industry
Technological Modernisation & Innovation	Driven by a need for advanced battery technology, charging infrastructure, and smart mobility solutions Promotes indigenous technological capabilities and cross-sectoral collaborations
Industrial Modernisation & Economic Restructuring	Transitioning traditional automotive manufacturing towards cleaner and more sustainable production processes Government incentives and subsidies for industrial upgradation
Environmental Sustainability & Ecological Civilisation	The development of the NEV industry offers a pathway to reduce greenhouse gas emissions, air pollution, and reliance on fossil fuels NEV Polciy clearly aligns with China's vision of ecological civilisation by promoting cleaner and greener mobility solutions
Socio-economic Development & Urbanisation	Promotes cleaner and greener urban mobility patterns and enhances the quality of life in urban metropolis Catalyses the job creation across the value chain of the New Energy Vehicles Industry, thus contributing to the broader objective of employment generation and inclusive growth

and socio-economic equity. In light of this, promoting New Energy Vehicles (NEVs) becomes a crucial tool for promoting China's SDG agenda and accelerating advancement in several sustainability-related areas.

China's engagement with the SDGs reflects a comprehensive approach to addressing pressing global challenges while advancing national development priorities. With a focus on poverty alleviation, environmental conservation, and inclusive growth, China has integrated the SDGs into its policy frameworks, aligning domestic strategies with international commitments. Key areas of emphasis include renewable energy deployment, pollution control, sustainable urbanisation, and climate resilience, all of which intersect with the NEV sector (see Table 11.2).

ECOLOGICAL MODERNISATION THEORY
AND NEW ENERGY VEHICLE INDUSTRY

In China, the concept of Ecological Modernisation Theory (EMT) has amassed significant attention and scrutiny among scholars, who advocate for its integration into the country's environmental policies and economic strategies (Kai-tai, 2006). Despite criticisms labeling it as merely "greening capitalism," EMT has demonstrated progress in the European Union and is gradually finding implementation within China's policy frameworks, albeit with acknowledgment of the time required to

Table 11.2 Here is a table about how NEV Policies are linked with SDG Goals in China

SDGs	Description	NEV Policy Impact
Goal 7: Affordable and Clean Energy	Ensure access to affordable, reliable, sustainable and clean energy for all	Reduce reliance on fossil fuels for transportation Increase energy efficiency and reduce carbon emissions through incentives for electric vehicles
Goal 9: Industry, Innovation and Infrastructure	Build resilient infrastructure, promote inclusive and sustainable industrialisation and foster innovation.	Drive advancements in battery technology, electric drivetrains, and smart mobility solutions Strengthen technological capabilities and enhance industrial competitiveness through R&D and collaboration between national and international stakeholders
Goal 11: Sustainable Cities and Communities	Make cities and human settlements inclusive, safe, resilient and sustainable	Mitigate urban air pollution and greenhouse gas emissions Promotes sustainable urban mobility and enhances livability through investments in charging infrastructure and public transportation electrification
Goal 13: Climate Action	Take urgent action to combat climate change and its impacts	Decarbonises the transportation sector, a major source of emissions Spending and investments under various NEV policies in China align with the country's NDCs (Nationally Determined Contributions for Sustainable Development) under the Paris Agreement to reduce carbon intensity and increase non-fossil energy sources

realise its full potential. This theory, which underscores the importance of achieving a harmonious balance between economic growth and ecological sustainability, aligns closely with China's strategic development objectives. An exemplification of this alignment can be observed in China's ambitious electric vehicle (EV) policies, where the nation aims to tackle environmental challenges while fostering technological innovation and economic growth. Through stringent regulations, financial incentives, and technological advancements, China is driving the transition towards electric mobility, illustrating a proactive approach towards ecological modernisation. For instance, implementing emission standards and quotas for electric vehicles, alongside substantial subsidies and tax breaks for EV purchasers, showcases the government's commitment to incentivising green practices in the automotive industry. Additionally, China's strides in battery technology, exemplified by the dominance of domestic firms in the global electric vehicle battery market, underscore its dedication to technological innovation in pursuit of environmental sustainability. These examples underscore ecological modernisation theory's relevance and practical application within China's evolving environmental and economic landscape.

CONCLUSION

In conclusion, the exploration of China's New Energy Vehicles (NEVs) industry and its alignment with the philosophy of ecological civilisation underscore a transformative journey towards sustainable development. China's commitment to ecological civilisation, exemplified by President Hu Jintao's 2007 report to the 17th National Party Congress, serves as a cornerstone for transformative change. This commitment integrates various components crucial for green development, emphasising the harmonious relationship between human activities and the environment.

The NEV industry, as a pivotal component of China's green development trajectory, has witnessed exponential growth and ambitious targets outlined in the New Energy Automobile Industry Development Plan (2021–2035). Key objectives include achieving fuel economy targets and market share goals for NEVs, emphasising technological innovation, structural adjustment, and sustainable resource management.

The state-led initiatives in the NEV sector are reshaping the socio-economic landscape of green mobility, reflecting a blend of economic growth and environmental stewardship. By prioritising sustainable practices, fostering innovation, and emphasising public engagement, China is poised to lead the transition towards a greener automotive future while advancing its vision of ecological consciousness and green development.

The implications of China's NEV industry development policy extend beyond its borders, positioning China as a global leader in green development and the automotive industry. However, further research is warranted to delve deeper into the nuanced interplay between policy deployment, ecological civilisation, and the narrative of sustainability versus development in shaping China's socio-economic trajectory and its role in the global arena.

In summary, the nexus between China's NEV initiatives and ecological civilisation highlights a paradigm shift towards a more sustainable and harmonious relationship between human activities and the environment, with profound implications for China's domestic development and its global leadership in green technology and innovation.

REFERENCES

Asian Development Bank (2012). *Towards and Environmentally Sustainable Future: Country Environmental Analysis of People's Republic of China.* Asian Development Bank. Manila. https://www.adb.org/sites/default/files/public ation/29943/toward-environmentally-sustainable-future-prc.pdf

Chenggang, X. (2011). The fundamental institutions of China's reforms and development. *Journal of Economic Literature, 49*(4), 1076–1151.

China Power Team. (2016, February 15). How is China's energy footprint changing? *China Power.*

Deng, T. (2017, June 2). In China, the water you drink is as dangerous as the air you breathe. *The Guardian.*

Economy, E. (2007, September/October). The great leap backward? *Foreign Affairs.*

EVBoosters. (2024, September 27). China exceeds 1 million ev sales in August 2024, setting a new record. https://evboosters.com/ev-charging-news/china-exceeds-1-million-ev-sales-in-august-2024-setting-a-newrecord/#::text=China's%20leadership%20in%20the%20global,are%20expanding%20their%20global%20presence

Feng, Suling & Chong, Yu & Li, Guoxiang & Zhang, Shubo. (2022). Digital finance and innovation inequality: Evidence from green technological innovation in China. *Environmental Science and Pollution Research. 29.* https://doi.org/10.1007/s11356-022-21826-2.

Full Text of Hu Jintao's Report at 17th Party Congress. (2007, October 15). National congress of the communist party of China. Retrieved February 5, 2024, from https://np.china-embassy.gov.cn/eng/Features/200711/t20071104_1579245.htm

Guoxia, M., Peng, F., Yang, W., Yan, G., Gao, S., Zhou, X., Qi, J., Cao, D., Zhao, Y., Pan, W., Jiang, H., Jing, H., Dong, G., Gao, M., Zhou, J., Yu, F., & Wang, J. (2020). The valuation of China's environmental degradation from 2004 to 2017. *Environmental Science and Ecotechnology, 1,* 100016, ISSN 2666-4984, https://doi.org/10.1016/j.ese.2020.100016. (https://www.sciencedirect.com/science/article/pii/S2666498420300089)

Hu, A., Tang, X., & Yan, Y. (2018). *Xi Jinping's new development philosophy.* Springer Singapore.

International Monetary Fund. (2020). World Economic Outlook, October 2020: A long and difficult assent, Washington, D.C.

Jahiel, A. R. (1998). The organization of environmental protection in China. *The China Quarterly, 156,* 757–787.

Kai-tai, Y. (2006). Summary of theories on ecological modernization. *Teaching and Research.*

Kendall, M. (2018). Fuel cell development for new energy vehicles (NEVs) and clean air in China. *Progress in Natural Science: Materials International, 28*(2), 113–120.

Li, Yang, et al. (2022). Dynamic pattern and evolution trend of the new four modernizations synchronous development in China: An analysis based on panel data from 31 provinces. *Sustainability, 15*(8), 6745, https://doi.org/10.3390/su15086745. Accessed 27 Mar. 2024.

Liangyu. (Ed.). (2018, May 20). Xi vows tough battle against pollution to boost ecological advancement. Xinhua Net. https://www.xinhuanet.com/english/2018-05/20/c_137191762.htm#:~:text=Xi%20vows%20tough%20battle%20against%20pollution%20to%20boost%20ecological%20advancement,-Source:%20Xinhua%7C%202018&text=BEIJING%2C%20May%2019%20(Xinhua),on%20environmental%20protection%20ending%20Saturday.

Lo, K. (2015). How authoritarian is the environmental governance of China? *Environmental Science and Policy, 54,* 152–159. https://doi.org/10.1016/j.envsci.2015.06.001

Ministry of Ecology and Environment of Peoples' Republic of China. (2018, February 13). The 2018 National Working Conference on Environmental Protection held in Beijing. Retrieved February 1, 2024, from https://english.mee.gov.cn/About_SEPA/leaders_of_mep/liganjie/Activities_lgj/201802/t20180213_431467.shtml

Peksen, M. (2021). Hydrogen technology towards the solution of environment-friendly new energy vehicles. *Energies, 14*(16), 4892. https://doi.org/10.3390/en14164892

Permanent Mission of the People's Republic of China to the United Nations Office at Geneva and Other Organizations in Switzerland. (1996). Environmental Protection in China, Information Office of the State Council of the People's Republic of China, Beijing.

Pruden, G. B. (1986). Economic Development in The People's Republic of China: Effects of The Four Modernizations. *Journal of Third World Studies, 3*(1), 36–43. https://www.jstor.org/stable/45197200

Shen, W., & Jiang, D. (2021). Making authoritarian environmentalism accountable? Understanding China's new reforms on environmental governance. *The Journal of Environment & Development, 30*(1), 41–67. https://doi.org/10.1177/1070496520961136

Stanway, D. (2019, April 16). China soil pollution efforts stymied by local governments: Greenpeace. *Reuters.*

State Council. (2020, October). Notice of the General Office of the State Council on the issuance of the new energy automotive industry development plan 2021–2035. In *State Council of the People's Republic of China.* State Council of the People's Republic of China. Retrieved March 5, 2024, from https://www.gov.cn/zhengce/content/2020-11/02/content_5556716.htm

Su, C., Yuan, X., Tao, R., & Umar, M. (2021). Can new energy vehicles help to achieve carbon neutrality targets? *Journal of Environmental Management, 297*, 113348. https://doi.org/10.1016/j.jenvman.2021.113348

Tang, B., Wu, X, & Zhang, X. (2011). Modeling the CO2 emission and the energy saved from new energy vehicles based on the logistic-curve," *Centre for Energy and Environment Policy Research*, Beijing Institute of Technology, [Working Paper 20]. https://ceep.bit.edu.cn/docs/2018-10/20181011134744377398.pdf

To, Y. (2024). *Contested development in China's transition to an innovation-driven economy.* Routledge.

U.S. Energy Information Administration. (2024, April 16). China imported record amounts of crude oil in 2023—U.S. Energy Information Administration (EIA). https://www.eia.gov/todayinenergy/detail.php?id=61843#:~:text=China%2C%20the%20world's%20largest%20importer,according%20to%20China%20customs%20data

Xiang, C., & Lo, A. Y. (2024). Authoritarian environmentalism 2.0: An incremental transition of environmental governance in China. *Environment and Planning C: Politics and Space.*https://doi.org/10.1177_23996544241286325

Xi, J. (2017). *The governance of China.* Foreign Languages Press.

Yu, R. (2016). China fines five auto makers for electric-vehicle subsidy fraud. *The Wall Street Journal*. https://www.wsj.com/articles/china-fines-five-auto-makers-for-electric-vehicle-subsidy-fraud-1473337367

Yuan, X., Liu, X., & Zuo, J. (2015). The development of new energy vehicles for a sustainable future: A review. *Renewable and Sustainable Energy Reviews, 42*, 298–305.

Zahoor, A., Yu, Y., Zhang, H., Nihed, B., Afrane, S., Peng, S., Sápi, A., Lin, C. J., & Mao, G. (2023). Can the new energy vehicles (NEVs) and power battery industry help China to meet the carbon neutrality goal before 2060? *Journal of Environmental Management, 336*, 117663. https://doi.org/10.1016/j.jenvman.2023.117663

Zhang, L., Mol, A. P., & Sonnenfeld, D. A (2007). The interpretation of ecological modernisation in China. *Environmental Politics, 16*(4), 659–668. https://doi.org/10.1080/09644010701419170

Zhang, Q., & Crooks, R. (2012). *Toward an environmentally sustainable future: Country environmental analysis of the people's Republic of China*. Asian Development Bank.

CHAPTER 12

India's Experiment with Electric Vehicles: Promise and Ambiguity in the Context of Environmental Security

S. Shekin and Haans J. Freddy

Abstract India is dedicated in reducing carbon emissions and reliance on traditional fuel sources, which has intricate interlinkages with national security. Electric Vehicles (EVs) are crucial in bridging nation's aspirations for a sustainable and technology-oriented future in the transportation sector. Promotion of EVs is the primary aspect of the nation's greening agenda. Convolutions from the production processes of EVs to meticulous assessments of their environmental ramifications throughout their lifecycle are addressed, examining their long-term durability and claim towards pollution-free transportation. India's pursuit to attain sustainable

S. Shekin (✉) · H. J. Freddy
Department of Political Science, Madras Christian College (Autonomous) (MCC), Chennai, Tamil Nadu, India
e-mail: shekinsancho3366@gmail.com

H. J. Freddy
e-mail: haansfreddy@mcc.edu.in

J. T. Karackattu et al. (eds.), *Environmental Securitisation in India and China*, https://doi.org/10.1007/978-981-97-9160-6_12

mobility is examined through the lens of Copenhagen School's securitisation theory. This finds relevance as India has ambitious goals in the environmental context. The advocacy of adoption towards EVs is promoted through policy frameworks, incentive programmes, subsidies and tax exemptions. This research will put forward arguments that there is an ambiguity that surrounds this initiative.

Keywords Electric vehicles · Sustainability · Securitisation · Green mobility · Ambiguity

INTRODUCTION

India is among the fastest-growing economies and has the potential to have the largest population in the world, in the near future. The major transition of energy in India in terms of shift from the conventional source and the greening, where environmental considerations are brought into the affairs of policymaking of the state happens to be the electrification of the transport sector. The transport sector in India has had a huge transformation in the recent years. While road transport expansion and improvement typically serve as a catalyst for socio-economic development, as has happened in many countries, it has unleashed several negative environmental problems in India, namely, burgeoning emissions of CO_2 as well as air pollutants such as nitrogen oxides (NO_X) and fine particulate matter ($PM_{2.5}$).(International Energy Agency) Currently, the Government of India has been trying to reduce the emissions of the carbon elements and other harmful polluting threats to the environment. For these initiatives have been taken through means of policies, action plans and legislations. India has brought in several ambitious policies to reduce the net energy demand by around 30% and also avoid the 60% of the expected carbon emissions by 2050. The visions of India's greening initiatives are well clear through their strategies. In 2021, India's Prime Minister had announced India's ambition to reach net-zero carbon emissions by 2070, at the 26th session of the United Nations Framework Convention on Climate Change (COP 26) (Forest & Climate Change. "Net Zero Emissions Target", 2023).

Is it the environmental aspects, the only concern for the huge efforts and multiple initiatives taken by a country to have an energy transition, being the second-most populous country in the world? No, the aspect of energy security also plays a major role. Energy security is vital for economic, political and social stability of India. Some countries with lesser resources tend to be dependent on other countries to secure their energy security. Some of the important elements of the energy security include maintaining a strong energy infrastructure, maintaining sustainable supply of energy through means of international treaties and cooperations, promoting the efficient use of the energy resources and diversifying the energy sources to ensure survivability in times of crises. In order to avoid any kind of fluctuations in the market, it is the responsibility of a country to ensure that there is reliable and stable access to the energy resources, in an affordable manner to meet the current and future needs of the country. Energy security can be defined as the ability of a nation to engage in strategic activities to meet the demands of the market by procuring and supplying the energy and thus ensuring stable and affordable energy supply for the development of the nation by minimising the vulnerabilities in energy chain. This could possibly ensure the national security of any country, as the domestic market demands are met without any tensions. Essentially, energy security comprises elements of national security, economic security and environmental security. Unlike some other forms of security, it is more 'vulnerability-based' rather than 'threat-based' (Rumley & Chaturvedi, 2015).

The conventional mode of energy used in the transport sector is oil. And India is the second-largest importer of crude oil in the world. India depends on foreign markets for 87.4% of the total crude oil imports, for their domestic use (www.ETEnergyworld.com, 2023). This is a major threat to the national security of the country. It is not at all just an aspect of energy security, as any geopolitical tensions or challenges have the potential to directly hit the Indian economy. And hence for this purpose, several key strategies such as diversification, diplomatic ties, energy dialogues with oil exporting countries are all done. The primary goal was to ensure the national security that has a potential threat through the heavy imports of the commodity of oil. This is also one of the driving sources for the energy transition from the conventional oil-based energy sector to the electricity-based energy sector in India. India, is on the verge of becoming a manufacturing hub for the electric vehicles industry. The Government of India aims to achieve 100% local production of electric vehicles under its 'Make in India' initiative (Jain, 2023). But the electric vehicles manufacturing industry is still only in very early stages, but with a quick growth plan.

The Process of Greening

The concept of greening is added to the policymaking process. This means that, environmental considerations and aspects of conservation of ecosystem is prioritised during the process of policymaking. They are all done for the purpose of sustainable development. The vagueness of 'sustainable development', a weakness in terms of technical discourse, gains a certain political strength because it allows political actors 'to proceed without having to agree also on exactly what to do' (ibid., p. 168; cf. O'Riordan, 1998; Hajer, 1997, p. 30). The idea of sustainable growth and development has not been clearly defined, where they are mentioned and the major drawback is that, they are perceived in different ways at different places. Whatever might be the theoretical perspective of the term 'sustainable development', but the common perceived idea of sustainable development is the development that meets the needs of the present without compromising the ability of future generations to meet their own needs (Development, 2023). India in the past two decades have evolved its legislation to accommodate the aspect of greening into it. Most of the policies and legislations brought in by the Government of India has a study on the environmental impacts brought in by making it come into force. The impact on the surrounding environment is well studied and determined before the legislation comes into force. This shows India's commitment to conservation of environment through process of greening. Though the intention is to have a green growth forward, the methods adopted by the government must be reconsidered in terms of the long-term effects.

One of the major initiatives brought in by the Government of India, happens to be the promotion of electric vehicles in India, especially in the passenger automobile segment. An electric vehicle (EV) is one that operates on an electric motor, instead of an internal-combustion engine (ICE) that generates power by burning a mix of fuel and gases (Standard, 2012). These vehicles are seen as possible replacements for the current generation conventional automobiles, to address various issues of pollution, global warming, depletion of natural resources, etc. The concept of EVs has drawn in more interest in the past decade, due to the rising carbon footprint and other environmental impacts of fuel-based vehicles. Electric mobility has been growing in a very rapid phase throughout the world. As people got to know about the impact of the traditional fossil fuel vehicles on environment, and think that electric vehicles are a cleaner and sustainable alternative and then shift to modern electric vehicles as there is no exhaust pollution. The early EVs first came into existence in the late nineteenth century, when the Second Industrial Revolution brought in the electrification of several sectors (Contributors, 2019). The major reason for people being part of energy transition to electric mobility is the concern over fuel prices and depletion of fossil fuels, and these vehicles are promoted with the tagline of being environmentally friendly. But the real factor of being eco-friendly has to be examined from the manufacturing process to their disposing, after the end of their usage completely.

India has set up huge aspirations in achieving their energy and sustainability goals, through EVs. The Government of India promotes the electric mobility as the advanced technology-oriented future of the transportation sector. In the 2021 United Nations Climate Change Conference (COP 26), India has unveiled its ambitious decarbonisation target for 2030. In the 2023–24 Union Budget, Indian Finance Minister Mrs. Nirmala Sitaraman had announced a budget allocation of INR 35,000 crore for crucial capacity investments aimed at achieving energy transition and net-zero targets by 2070 (Singh, 2023). Most of the carbon emissions are contributed by the transport sector. There must be some kind of changes and alternatives on the existing traditional methods of transport, to achieve the environmental goals. But there are lot of challenges as India is a huge consumer market and third-largest automobile market in the world. But there is a possibility as the automotive sector is still in the early growing stages and the high number of consumers happen to be due to the huge population of India. In 2023, EVs had a market share of 6.3%, which is double the time of the previous year. India tries to promote itself as the major transitioner in the automobile and energy sector. And the issue of national security is yet another major aspect, which has a threat from the oil supply disruptions and price volatility. This is also a reason for the energy transition towards ramified electrification of domestic mobility. India is among the few countries that currently support the global EV30@30 campaign, which aims for at least 30% new vehicle sales to be electric by 2030 (https://www.drishtiias.com/daily-news-editorials/electric-vehicles-india-s-future).

Indian government in their official pages, have been promoting the EVs as a cleaner and greener mode of transport. The main focus of their promotion is based on the low running cost. The government has been working towards the benefits in charging also, provided this would turn the running cost to negligible amount. India aims to find an alternate to fossil fuel-based transport sector, as India has to depend on foreign markets for the fuel. And India's pursuit to attain an alternate to ICE vehicles have been only with EVs and no other hydrogen or ethanol-powered vehicles have been promoted. EVs are considered as the future of Indian mobility by the Indian government. There are not only EVs in the private sector, even the public vehicles are also being transitioned to EVs. The idea of the government is to bring in good industry for the country by promoting the products within their territory first. When there is a huge market potential enough to make profit, it is more likely to pull in more investors from the international level, which is healthier for the Indian Economy. And once there is a potential infrastructure for the manufacture and consumption of EVs, India would become a manufacturing hub for the EV sector. This would give more powers to India, as many countries would depend on India as they give supply of goods. The overall imperative is clear that, the promotion of EVs in the Indian market, at the domestic level is not just the environmental perspective but also has aspects of the power dynamics of energy and trade sector too.

At the global level, the EV sector had already had a huge impact on the automobile markets. The governments have been facilitating the shift to the modern electric energy. And EV manufacturing companies have been on a rise for more than a decade. In the global level, it's the cars that have a huge production in EV sector. Only in the Asian regions, the electric mobilisation of the two wheelers have been popular. The EV industry is new to the market and many investors have already taken huge sums of returns, post investments. The new industry had a very high potential of giving out more returns to the investors. This brought in interest between the global actors and giant investors to get hold of the industry, and potentially grow along with the industry. This, over a period of decade had created a great positive outcome. And the countries which are part of European Union (EU) have also been keen in bringing in EVs into their market as an obligation by bringing in ban on sale of ICE cars and also scrapping of existing ICE vehicles in a given time limit. This has brought in a great increase in the market share of EVs, in the Western markets. But the advantage of the Western world is that, they have the technological advancements and the infrastructure enough to support the electric mobility. The countries in the European region have enough of the raw materials to build the EVs, such as Lithium, Cobalt and Nickel (commonly known as rare earth metals). And then the infrastructure required for its running and the battery replacement mechanisms, as the batteries used in the EVs are by nature, depleting with time.

The major motivating force happens to be the sanctions imposed on Russia as a response against the Russia-Ukraine war. Due to the sanctions the EU countries could not get oil from Russia and hence, they were forced to buy oil from the Middle East for a very higher amount of price. In a period of year, the oil prices shot up to twice the amount. These geopolitical tensions had affected those countries a lot. The sanctions were taken too serious by the members of EU, as they had also imposed sanction on the finished petroleum products that was exported from India to EU, as they were keen in banning the products made from Russian oil. As far as EU is concerned, they have good sources of generating electricity from clean renewable sources. And hence, it is viable to have EVs as an alternate to the ICE vehicles. But India aims to replicate the models of various countries and also relies on with high ambitions and having a great positive output. But how does India generate electricity? 75% of India's electricity is generated from coal-powered thermal power plants (Government of India, 2024), which is hazardous to the environment as it emits lot of pollutants. And hence, it can't be green to run EVs in India, as the whole of electricity generated itself is not from clean source. India must focus on more of renewable sources of energy, to meet their energy goals. In a country like Norway, where most of the electricity is produced from hydropower, it is comparatively cleaner and makes more sense from EVs in their market.

However, the environmental concerns in the aspect of EVs starts right from the manufacturing of their batteries. The production and disposing of EV batteries pose serious environmental challenges. Lithium and cobalt are the very essential rare earth metals, needed for lithium-ion battery production. The cobalt which helps power them and the other technologies comes with some serious humanitarian concerns, where it is mined in the Democratic Republic of Congo (DRC) (Cobalt powers our lives. What is it—and why is it so controversial (2023). The United Nations have cited that DRC's 'deteriorating security situation', its humanitarian crisis affecting 27 million people, as well as child-labour practices and the ongoing guerrilla campaign waged over the exploitation of resources and food security (Evans, 2022). The humanitarian crisis is a major issue in the DRC, where women and children are also going through ill treatments during the extraction of these rare earth metals. The lithium and cobalt are then processed in stages and then sent to battery manufacturing industries. Though there are several players in the field of mining, China dominates material processing, cell component manufacturing, battery cells and overall EV manufacturing sector at the global level (IEA, n.d.). An average of 60% more carbon is emitted in producing an EV than a conventional fuel-powered vehicle. This is primarily due to the manufacturing of batteries. And it has to be noted that a normal fuel-powered car has approximately 2000 moving parts in the engine, whereas an electric car only has 20 (Idaho National Laboratory, 2014b). But one tonne of lithium to be extracted costs 5,00,000 litres of water and emits around 10 tonnes of carbon(Tedesco, 2023). The primary concern on sustainability can't be justified due to the overall carbon emissions in production process of battery itself.

It is very clear that, the mining of the rare earth materials such as lithium and cobalt, for the manufacturing of lithium-ion batteries is hazardous to the environment. But once the EVs are made, the greener use depends on the source from which the electricity is generated. For example, in a country like Norway, where most of the electricity is generated from renewable source of energy, the hydropower, it makes the electric mobility greener due to the lowest emissions of power sector across Europe. The country has 1769 hydropower plants (about 88% of Norwegian production capacity), 1240 storage reservoirs and 65 wind farms. This makes it more sustainable environment to develop the EV infrastructure. But India already generated 75% of its electricity needs through coal-powered thermal power plants (Government of India, n.d.). It is not environmentally friendly to drive an EV with electricity sourced from non-renewable ways. It possesses serious threat to the environment as same as conventional vehicles. The only difference is that, conventional gasoline-powered vehicles have exhaust emissions but EVs do not have that, but the impact on the environment with the carbon footprints are inevitable in both the cases. The batteries of EVs are also depleting in nature. In few years, or after thousands of miles, the battery tends to lose its capacity. The battery needs to be replaced for the further use of EVs or else there is no other solution rather than abandoning them. In both the cases, lithium-ion battery dumping is done and it is very hazardous to the environment. Only in the recent days, recycling them has been a developing concept. There are dangers of thermal runaway, where any contact with salt water and combustion happens in lithium-ion battery even without oxygen and outrageous fire is formed (Shahid & Agelin-Chaab, 2022). There need to proper recycling mechanisms for both EVs and batteries as they are harmful to the environment than conventional vehicles.

Despite all the negative and harmful consequences that EVs pose to the environment and the consumer, The Government of India seems to be pushing for the enlargement of EV production with serious intent, the reasons for which are not very clear.The Automotive Research Association of India (ARAI) is an autonomous body affiliated to the Ministry of Heavy Industries, Government of India. This body is responsible for the testing of automobiles and the related prospects for the benefit of customer. It is stated that, the tests for the range of EVs are done by ARAI in closed environment, at ideal conditions in the absence of any resistance. This could give ideal and high range on papers. This misleads the consumers as they decide on their daily commute and finally the range anxiety is also made. There are lot of consumer cases, as the EVs are not delivering the range as provided by the ARAI. But this relates to another problem. It is the lacking of charging infrastructures throughout India. The sale of EVs is ramping up, but there are no proper charging facilities, which is worrisome as without charge, the commuters have no option other than being stranded and towed to the next charging station. The current ration is approximately 1 charging station per 135 EVs, which is significantly lower than the global ratio of 1 charging station per 6 to 20 EVs and this shortage in charging stations has the potential to push India to be 40% behind its EV30@30 vision (Bharadwaj, 2023).

India's Ambitious Goals and Sustainable Mobility

India is being very positive in achieving sustainable development and environment security through their electric mobility goals. But there are serious harmful hazards to the environment in the case of India. India's goal of achieving net-zero greenhouse gas (GHG) emissions by 2070 was a much-discussed element of the national statement at COP26, as India was very few countries in G20, that had not committed to a net-zero target before GlasgowCGEP, 2021). It is good that India has ambitious goals towards a sustainable future. But is electric mobility a feasible way to attain it is still a paradox. It revolves around various factors and India should focus more on clean energy transition before electric mobility. The Paris Agreement and National Action Plan on Climate Change (NAPCC) are some of the steps taken for this process. Yet, it is done in a slower place compared to electrification of transport sector. Some of the major initiatives taken to promote the electric mobility sector are the Faster Adoption and Manufacturing of Hybrid and Electric Vehicles (FAME) schemes (Fame India Scheme, 2023a), which provide financial incentives to manufacturers and buyers of EVs. In some aspects, they also tend to improve the charging infrastructure in India. Even the individuals who purchase EVs can evade income tax by a good per cent on the interest paid on EV loans. During the initial days, many charging stations that were run by the Government was run at free of cost. This means, any EV owner can drive into a charging station and plug it until, the EV is completely charged and then leave without paying a dollar. This had attracted lot of people into the EV segment.

But apart from this, the Government of India's website all promoted electric mobility as the cleanest mode of transport and it is the only way to achieve net-zero carbon emission targets. Most of the information given in the official websites of the government seems to be not acceptable in nature. But they happen to be more attractive to the consumer as they provide purchase incentives, coupons, interest subventions, road tax exemptions, registration fee exemption and even income tax benefits (Government of India, n.d.-a). But the long-term effects of introducing a new big sector in the automobile industry and the potential threat it has in the future is inevitable. The primary objective of the initiatives taken by the Government of India were to boost the domestic manufacturing capacity for EVs. This was the initial steps taken in order to make India a manufacturing hub for the EV sector. The second objective was to create more job opportunities in the EV industry. And thus, promotion of entrepreneurship and foreign investments and setup are encouraged in this perspective. This has the heavy potential to boost the Indian Economy too. The third objective was to reduce the dependence on imports with regard to EVs and their components. For this, the Government required huge support from the domestic level too. And to enable the trusts of the public, the vehicles in the public transport system such as buses and taxis have been converted into electric mobility slowly. States like Kerala added large fleets of EVs to their Motor Vehicle Department (MVD), which can bring about huge change in the minds of people. These are kinds of soft power methods used.

The Copenhagen School of Securitisation is very important aspect of the security studies, as an act of speech than an objective threat. The state actor must label a potential threat or an issue as an existential crisis, by which extraordinary measures have to be legitimised and the activities can move beyond the realm of normal politics. The Copenhagen School also emphasises on the social construction of security, by highlighting the importance of perceptions and interpretations to understand a threat or an issue. And this idea of securitisation is very much important for the Indian context, as there is a possibility of huge threat in the near future, if the situations are not dealt in a proper manner. Until now, there are no proper infrastructure for the battery replacement or recycling of used lithium-ion batteries of EVs. They can cause huge hazards to the environment post depletion and can even move towards an environmental crisis. Just because, the technologies are introduced in some modern markets with different conditions of energy sourcing, it necessarily need not be the solution for India too. This might be a boon in the context of several countries, but there is a severe threat of EVs to turn out to be a bane in the Indian transport sector. Without proper infrastructural developments that cater to the needs of the market, it makes less sense to promote the industry just for the beneficial aspects.

The situation can turn out to be volatile at any moment, as the domestic markets are not yet adapted to the energy transition. If there is a sudden impact on the sector, there is a possibility for a collapse in the overall structure. The consumer base will be the most affected and there will be huge distress created among the customer base, who had shifted into EVs. The government will also have a huge impact, as they would have included large fleets of EVs into the public transport sector. The time to realise would be too late and there will be no other alternatives left than coming back to the conventional fuel-based vehicles. There are only few research and development studies done in the EVs in majority sectors as of now. The actors have no or very few aspects regarding the long term. The sector had been of great advantage to the capitalists and the investors who had taken part in the boom of the EV sector and had made huge business based on their skills. India has more than 100+ electric two-wheeler sellers in India. Most of them don't even own a manufacturing hub. The process starts with importing the motorcycle from China, rebadging it and finally selling the product in the Indian markets. The service backup, long-term run and the disposal or battery change remains as a paradox.

It is not wrong for a government to promote a new sector or an alternative sector for the conventional industry. But in the context of EVs, the government should try to bring in better ways to produce electricity, which in turn will make the electric mobility a green method of transport. The battery replacement system and the proper mechanisms for the disposal of lithium-ion batteries must be brought in by the government. This would avoid further confusions on the long-term volatility of the market. There requires a transition between gasoline-based vehicles to EVs. The government can consider hybrid vehicles for a stage of transition, as they also contain lithium-ion batteries and this can be used as a testing mechanism for the full transition towards EVs. It is the responsibility of the government to bring in more awareness about the real idea of EVs that helps the consumers to understand the reality of the new sector. More importantly, proper charging infrastructures must be developed throughout the country, as there is more demand even now where the sector is only at the starting stage. And it is also necessary that proper assistance and solutions reach even to the remote places as many EV owners might even be at those places. The redressing of their grievances is also very important, as the state actor who had played a huge role in the promotion of a new sector, which had a huge role in the decision-making process of people in moving towards newer technologies. If only the environmental aspects are not looked into properly, there is a huge threat for the environmental security of India which possesses direct attacks on the national security of the country. Decisions taken at the earliest stages can save the country from crisis, before the ramping up of complete electrification process.

References

Bharadwaj, R. (2023, June 20). *EV Infrastructure in India: What to Expect By 2030 | Bolt.Earth Blog.* Bolt Earth. https://bolt.earth/blog/Indian-ev-charging-infrastructure-by-2030

Business Standard. (2012). *What is electric vehicle | Types of Electric Vehicles | Electric Vehicle News.* Business Standard. https://www.business-standard.com/about/what-is-electric-vehicle

CGEP, C.|. (2021, November 17). *Evaluating India's COP26 Climate Commitments|Q&A with Dr. Kaushik Deb and Mahak Agrawal.* Center on Global Energy Policy at Columbia University SIPA|CGEP. https://www.energypolicy.columbia.edu/publications/evaluating-india-s-cop26-climate-commitments-qa-dr-kaushik-deb-and-mahak-agrawal/#:~:text=India

Cobalt powers our lives. What is it—and why is it so controversial? (2023, December 21). Environment. https://www.nationalgeographic.com/enviro nment/article/cobalt-mining-congo-batteries-electric-vehicles#:~:text=The% 20silvery%20blue%20metal%20is

Evans, G. (2022, October 31). *Sharing insights elevates their impact.* S&P Global. https://www.spglobal.com/mobility/en/research-analysis/a-reckon ing-for-ev-battery-raw-materials.html

Government of India. (2023a, July 23). *Fame India Scheme.* Pib.gov.in.https:// pib.gov.in/PressReleaseIframePage.aspx?PRID=1942506#:~:text=This%20p hase%20mainly%20focuses%20on

Government of India. (n.d.). *Ministry of Coal, Government of India.* www. coal.nic.in. Retrieved February 20, 2024, from https://www.coal.nic.in/en/ major-statistics/generation-of-thermal-power-from-raw-coal#:~:text=In%20I ndia%2C%20power%20is%20generated

Government of India. (n.d.-a). *ELECTRIC VEHICLE INCENTIVES.* E-Amrit. niti.gov.in.https://e-amrit.niti.gov.in/electric-vehicle-incentives

https://www.drishtiias.com/daily-news-editorials/electric-vehicles-india-s-future

Hajer, Maarten A. (1997) *'The Politics of Environmental Discourse: Ecological Modernization and the Policy Process'*, Oxford: Oxford University Press.

Idaho National Laboratory. (2014b). *How do gasoline & electric vehicles compare?* Advanced vehicle testing activity. https://avt.inl.gov/sites/default/files/pdf/ fsev/compare.pdf

IEA. (n.d.). *Global supply chains of EV batteries.* https://iea.blob.core.windows. net/assets/961cfc6c-6a8c-42bb-a3ef-57f3657b7aca/GlobalSupplyChainsofE VBatteries.pdf

International Energy Agency. Executive Summary – Transitioning India's Road Transport Sector – Analysis. *IEA.* www.iea.org/reports/transitioning-indias- road-transport-sector/executive-summary

Jain, A. (2023, June 29). *How India accelerates towards becoming next power- house in EV production?* Mint. https://www.livemint.com/news/india/ev- industry-in-india-how-india-accelerates-towards-becoming-next-powerhouse- in-ev-production-11688005217114.html

Ministry of Environment, Forest and Climate Change. "Net Zero Emissions Target." *Pib.gov.in,* 3 Aug. 2023. pib.gov.in/PressReleaseIframePage.aspx? PRID=1945472#:~:text=India%2C%20at%20the%2026th%20session

O'Riordan, Timothy (1998) The Politics of Sustainability, in Turner, Robert Kerry (ed) *Sustainable Environmental Management: Principles and Practice,* Colorado: Westview Press. pp. 29–50.

Rumley, D., & Chaturvedi, S. (2015) (eds.). *Energy security and the Indian ocean region* (pp. 19–20). Routledge.

Shahid, S., & Agelin-Chaab, M. (2022). A review of thermal runaway preven- tion and mitigation strategies for lithium-ion batteries. *Energy Conversion and Management: X, 16*, 100310. https://doi.org/10.1016/j.ecmx.2022.100310

Singh, A. (2023, February 7). India's EV Economy: The Future of Automotive Transportation. www.investindia.gov.in. https://www.investindia.gov.in/team-india-blogs/indias-ev-economy-future-automotive-transportation

Sustainable Development. (2023). International EDS. Institute for Sustainable Development. https://www.iisd.org/mission-and-goals/sustainable-development#:~:text=Sustainable%20development%20has%20been%20defined

Tedesco, M. (2023, January 18). *The Paradox of Lithium*. State of the Planet. https://news.climate.columbia.edu/2023/01/18/the-paradox-of-lithium/#:~:text=To%20extract%20one%20ton%20of

Wikipedia Contributors. (2019, February 7). *Electric vehicle*. Wikipedia; Wikimedia Foundation. https://en.wikipedia.org/wiki/Electric_vehicle.5

www.ETEnergyworld.com. "India's Import Dependence on Crude Oil Climbs to 87.8% in April- August 2023: PPAC - et EnergyWorld." *ETEnergyworld.com*, 11 Sept. 2023, energy.economictimes.indiatimes.com/news/oil-and-gas/indias-import-dependence-on-crude-   oil-climbs-to-87–8-in-april-august-2023-ppac/103822860#:~:text=According%20to%20the%20PPAC%20data. Accessed 19 Feb. 2024.

Index

A

Aamra, 89
Accountability, 64, 166, 170, 172, 173
Acharya, Amitav, 5, 82, 86, 87
Actors, 4–7, 9, 14, 15, 17, 19, 30–34, 36, 37, 49, 53, 56, 61, 63, 66, 75, 76, 87, 98, 128, 131, 138, 146, 147, 162, 165, 168, 204, 206, 211, 212
Adani group, 52
Administration, 12, 98, 104, 109, 149
Advanced social and economic development zones (ASEZs), 117, 124
Agenda 21, 187
Agreement, 14, 31, 37, 39, 42, 62, 71, 72, 99, 101, 118–120, 151
Ahistoricism, 86
Alliances, 61, 66, 67, 69, 70, 131, 152
Amur
  Amur Region, 116
  Amur tiger, 119

Amursk, 120
Anarchophilia, 86
Angeeche, 89
Angrim Valley, 89
Anti-terrorist laws, 170
APEC, 154
Apiya, 89
Aquatic bioresources, 119
AR5 Synthesis Report, 156, 158
Aralia, 123
Arbitrary, 83, 172
Arunachal Pradesh, 89
ASEAN, 154
Assam, 48, 90
Atal Mission for Rejuvenation and Urban Transformation (AMRUT), 51
Authoritarian environmentalism, 184, 187, 188
Authoritarianism, 184
Automobile industry, 189, 210
Automotive Research Association of India (ARAI), 209
Auto sector, 189

B

Bear, 123
Beck, Ulrich, 55
Belt and Road Initiative (BRI), 54,
    136
Bewick, Thomas, 90, 91
Bijli Bachao Desh Banao, 150
BIMSTEC, 151
Biodiversity/bio-diversity, 7, 68,
    70–73, 75, 150, 154
Bodo, 89, 90
Border, 5, 66, 89, 106, 116, 119,
    121, 123, 124, 130, 172, 197
Border rivers, 119
Brandis, Dietrich, 92
BRICS, 138, 151, 154
BRI Ecological and Environmental
    Protection Big Data Service
    Platform, 136
BRI International Alliance for Green
    Development, 136
Brundtland Report, 129
Buzan, Barry, 5–7, 32, 33, 85–87,
    98, 102, 107, 128, 129, 137,
    138, 147, 150, 151

C

Carbon emissions, 38, 67, 68, 72, 98,
    151, 153, 154, 191, 202, 205,
    208, 210
Carbon neutrality, 154, 191
Carbon trades, 53
Central American, 92
Central National Security Commission
    (CNSC), 104
Centre for Science and Environment
    (CSE), 10
Char-dham, 48
Chemical waste, 186
Chennai, 48, 172
China, 5–15, 17–19, 30–32, 36–44,
    52, 53, 55, 99–106, 116–124,

128, 132, 134–136, 138, 139,
    149, 152–157, 162, 163,
    165–167, 170–174, 180–191,
    196, 197, 208, 211
Chinese, 8–11, 14–18, 54, 99, 101,
    102, 104, 106, 107, 116–120,
    122, 123, 134, 135, 152–157,
    165, 167, 172, 184, 186, 187,
    189, 190
China-Arab States Cooperation
    Forum, 136
China's Green Development in the
    New Era (CGD), 152, 153, 157
China's National Defence in New Era
    (CND), 152, 153, 157
Chinese Communist Party (CCP),
    135
Chinese Dream, 167, 183
Citizens, 8, 10, 15–17, 41, 74, 118,
    123, 150, 162, 165–169,
    171–174
Civilisationism, 136
Civilised, 137, 138
Civil society, 158, 163, 165, 166,
    168, 170, 173, 174
civil society organisation, 69
Class, 10, 163
Climate Action Network (CAN), 67
Climate agreements, 65, 106, 152
Climate change, 30–32, 34–44, 48,
    56, 60, 66–69, 72, 75, 76, 98,
    100–109, 129, 130, 133, 135,
    136, 148–158, 162, 180, 183,
    188, 190
Climate emergency, v
Climate policy/climate policies, 18,
    30–36, 41–44, 65, 69, 72, 73,
    76, 100, 101, 103
Climate policy advocacy, 70
Climate politics, 32, 36, 99–101, 103,
    108
Climate scientists, 65

Climate security, 36, 106, 107, 157, 158

Coalition for Disaster Resilient Infrastructure (CDRI), 133, 134

Cobalt, 53, 207, 208

Cold War, 4, 34, 41, 106, 108, 146

Collective action, 67, 68, 162, 167

Colonised, 88

Common but differentiated responsibilities (CBDR), 135

Communalisation, 49
    communities, 16, 18, 56, 63–67, 71, 88, 90, 92, 155, 156, 158, 162, 164, 166, 171, 173, 183

Communist Party of China (CPC), 8, 12, 13, 17, 165, 181
    Communist Party, 100, 102

Community, 6, 82, 83, 88, 101, 105, 107, 151–153, 156, 157, 166, 170, 171
    community-based decision making, 88

Compensation, 18

Compensatory Afforestation, 54

Compensatory Afforestation Fund Act (CAFA), 2016, 55

Compliance, 39, 84

Competitiveness, 74, 195

Conference of the Parties (COP), 67, 69, 71–73, 75

Congress
    18th Party Congress, 100
    19th Party Congress, 104, 105

Conservation, 6, 9, 14, 68, 70–73, 82, 88, 89, 105, 153, 182, 187, 204

Constructivist, 5, 6, 53, 56, 57, 60–64

Consumerism, 146, 188

Conventional vehicles, 208, 209, 211

Convention on Biological Diversity (CBD), 7, 39, 71

Convention on the Protection and Use of Transboundary Watercourses and International Lakes, 119

Copenhagen School, 6, 49, 53, 129, 131, 136–139, 146, 148, 211

COVID pandemic, 166, 172

Crab Eating Macaque, 54, 56

Critical school, 49, 50

Critical technologies, 191

Crude oil, 188, 189, 203

Cyber-crimes, 171

Cyclones, 48, 156

D

Danda, 83

Dasgupta, Chandrashekhar, 69

Decarbonisation, 38, 190

Deer, 89

Defence, 148, 149, 152, 157

Democracy/democracies, 8, 53, 148, 168, 172, 174

Democratic, 8, 17, 34, 61, 165, 172

Democratic Republic of Congo (DRC), 53, 207, 208

Demonetisation, 49

Desecuritisation, 137, 138

Desertification, 186

Developed nation, 102

Developing nation, 99, 100, 102, 106, 108, 157

Development
    development discourse, 35
    development policy, 180, 181, 197
    sustained development, 106

Dibang
    Dibang Tiger reserve, 89
    Dibang Wildlife Sanctuary, 89

Digital arena, 164

Digital finance, 189
    digital financial development, 190

Digitisation, 173
Diplomatic ties, 74, 203
Discourse power, 106, 135
DPRK, 119
Drought, 156

E
EAS, 154
Ecological civilisation, 100, 101, 105,
    180–183, 196, 197
    eco-civilisation, 152
Ecological consciousness, 180, 181,
    197
    eco-consciousness, 186
Ecological preservation, 183
Ecologisation, 49, 56
Ecology, 4, 11, 13, 14, 18, 52, 53,
    56, 105, 116, 119, 121, 124,
    181, 182, 184–186
Economic development, 4, 12, 30,
    34–37, 40–44, 99, 101, 118,
    122, 123, 136, 152, 157, 183,
    188, 190, 202
Economic growth, 8, 9, 12, 14, 15,
    17, 74, 100, 101, 104, 118, 122,
    153, 181, 184, 185, 188, 189,
    192, 197
Economisation, 52
Economy, Elizabeth, 186, 188
Ecosystem, 12, 16, 48, 52, 54,
    71–73, 75, 102, 119, 120, 124,
    204
Edwardian, 90
Efimov, Nikolai, 121
Electric energy, 206
Emission/emissions, 31, 37, 38, 40,
    42, 43, 68, 73, 98, 100, 101,
    103, 107, 109, 122, 133, 138,
    151, 190, 202, 208, 209
Endangered species, 71, 73
The Energy and Resources Institute
    (TERI), 37

*Energy-Saving and New Energy Vehicle
    Industry Plan*, 191
Energy security, 100, 106, 149, 188,
    189, 192, 203
Energy sources, 69
    energy power plant, 101
    solar, 101
    wind, 101
Environmental advocacy, 16, 164, 173
Environmental civilisation, 106
Environmental cross-border
    environmental cross-border
        cooperation, 117
Environmental degradation, 37, 75,
    98, 99, 118, 132, 134, 137, 139,
    184, 185, 187, 188, 190
Environmental diplomacy
    international environmental
        diplomacy, 68
Environmental Impact Assessments
    (EIAs), 54
Environmentalisation, 49
Environmentalising, 49
Environmental legislation, 119
Environmental police, 7, 100
Environmental policy/environmental
    policies, 4, 5, 8, 17, 19, 62, 73,
    116–119
Environmental politics, 98, 100
    international environmental politics,
        99, 108
Environmental protection, 37, 75, 98,
    100, 108, 116–119, 153, 167,
    174, 181, 183, 184, 187, 188
Environmental Protection Bureaus,
    100
Environmental Protection Law, 15,
    187
Environmental Protection Leadership
    Group, 187
Epidemiological diseases, 124
Epistemic, 65, 76

Epistemic communities, 60, 62–69, 71, 73, 75
Ethnocentrism, 86
Ethnographically, 48
Ethnographies, 18
Eurocentrism/eurocentric, 19, 86, 137, 139
European Union (EU), 42, 53, 98, 130, 206, 207
Evidence-based, 66, 69

F

Far East, 116, 117, 119, 121–124
Faster Adoption and Manufacturing of Hybrid and Electric Vehicles (FAME), 210
Fetishisation, 53, 56
Fifth estate, 163
Finnemore, Martha, 60–62
First India-Japan Environmental Policy Dialogue, 151
Fisheries, 119
Five-Point Agenda, 151
Five Year Plan
    13th Five Year Plan 2016–2020, 189
Floods, 48, 156, 172
Floyd, Rita, 8, 30, 31, 33, 36, 50, 131
Foreign investment, 74, 189, 210
Foreign market, 203, 206
Foreign policy, 33, 34, 39, 87, 100, 103, 104, 106
Forest fire, 119
Forest Rights Act (FRA), 2006, 16, 55
Forum on China-Africa Cooperation, 136
Foul feline, 90
Fourth Assessment Report, 98
France, 52
Free speech, 172

Functional actors, 7–9, 11, 17, 147
Fundamentals of the State Environmental Policy, 117

G

Ganga River, 150
Ganga Task Force, 150
GDP, 16, 101, 118, 122, 185
General Assembly, 155
Genetic resources, 71
Ginseng, 123
Glacier/glaciers, 108
Glasgow, 151, 209
Global environmental governance, 62, 63, 68, 183
Global EV30@30 campaign, 205
Global governance, 65, 66
Global North, 8, 39, 128, 130–132, 134, 137–139
Global politics, 32, 34, 55, 60, 62–64, 73, 75
Global South, 4–6, 18, 19, 39, 40, 75, 128–134, 136–139, 188
Global warming, 64, 67, 70, 74, 204
Global Witness, 53
Gobi Desert, 102
Good governance, 53, 132
Gramscian, 49
Greater Nicobar Island (GNI), 54
The Great Leap Backward, 185, 186
Green credit, 190
Green development, 101, 102, 151, 153, 180, 181, 189, 196, 197
Green finance, 136
Green fuel, 189
Green goals, 101
Green growth, 17, 118, 204
Greening, 5, 6, 9, 15, 17–19, 49, 50, 55–57, 73–75, 180, 202, 204
Green mobility, 180, 181, 192, 196
Greenpeace, 186

Green Strategic Partnerships, 151
Green, Susie, 90
Gross national product (GNP), 184
Gross regional product (GRP), 124
Guangdong, 185
Guideline/guidelines, 51, 109, 119, 169, 170
  green guidelines, 180
Gujarat, 48
Gujarat International Finance Tec-City (GIFT), 51
Guzzini, Stefano, 5

H
Haas, Peter, 62–66
Hansen, Lene, 162
Harbin, 120, 121
Harding, Richard, 92
Harmony/harmonious, 4, 5, 7–9, 12, 13, 15, 17, 53, 90, 101, 102, 105, 107, 151, 181–183, 186, 196, 197
Hazardous, 10, 118, 207, 208
Heavy metals, 186
Hebei, 185
Henan, 185
Himalayas, 48, 52
Honduras, 92
Huayuquan, 135
Human Development Report, 155
Humanitarian, 135, 152, 207, 208
Humanitarian assistance and disaster relief (HADR), 153
Human rights, 50, 61, 65, 135, 156
Human security, 9, 34, 107, 148, 153, 155–158, 166
Hurricane Katrina, 171
Hydrogen energy, 190
Hydrological, 50

I
Identities, 60, 61, 63, 73, 75
Ideology/ideologies, 13, 41, 85, 106, 109, 146, 180, 189
Indian Policy, 149
Indonesia, 55
Industrialisation, 85, 146, 188
Indus Valley Civilization, 88
Inequality, 18, 189, 190
Infrastructural communication, 170
Innovation, 13, 19, 66, 74, 181, 183, 189, 190, 196, 197
  green technological innovation, 15, 17, 19, 189
Inspector General of Forests, 92
Institute for Defence Studies and Analyses (IDSA), 149
Institutional relationships, 184
Intergovernmental Panel on Climate Change (IPCC), 98
Internal Combustion Engines (ICEs), 190–192
International, 5, 12, 14, 18, 19, 30, 32, 36–39, 41, 42, 53, 54, 60–63, 65–70, 72–76, 85, 87, 98–106, 108, 116, 118, 120, 121, 123, 128–131, 133, 135, 137–139, 146, 147, 151–154, 157, 162, 169, 187–189, 203, 206
International anarchy, 129
International cooperation, 74, 105, 183
International Monetary Fund (IMF), 184
International peace, 30, 35, 129–131, 134, 139
International relations, 60, 66
International relations (IR), 5, 6, 86–88, 92, 128, 136, 155
International Relations Theory (IRT) Non-Western IRT, 82, 92

Traditional IRT, 82, 92
International Solar Alliance (ISA), 32, 42, 70, 133, 134, 151
Irkutsk, 122

J
Jawaharlal Nehru National Urban Renewal Mission (JNNURM), 51
Jewish Autonomous Region, 116, 122
Jiangsu, 185
Jilin, 120
Jinping, Xi, 12, 15, 98, 99, 101, 104, 107, 109, 152, 154, 167, 181–183
Jintao, Hu, 12, 13, 100, 109, 180, 196
Journalists, 174
Justice, 16, 17, 134

K
Kashmir, 172
Kaviraj, Sudipta, 82, 83, 85, 92
Keqiang, Li, 12, 104
Kerala, 48, 88, 210
Khabarovsk, 120
Khan, Sher, 91, 92
Kipling, Rudyard, 91
Kunming, 154
Kyoto Protocols, 37–40, 100

L
Lakes, 183
Land degradation, 186, 187
Legitimation, 55, 84
Leopard, 123
Lifestyle for Environment (LiFE), 151
Lithium, 53, 207–209, 211, 212
Little, Arthur D., 163
Little, Richard, 86, 87
Livelihood, 155, 156, 170

M
Macferson Lake, 150
Machans, 90
Macropoliticisation, 102, 107
Macrosecuritisation, 98, 102, 107, 108
Manas Tiger Conservation Program, 89
Manniu River, 121
Mao
  Mao Zedong, 99
  post-Mao, 100
Marginalisation, 16, 134, 166, 170, 173
Maritime, 101
Market access for technology, 189
Market segmentation, 189
Marxist, 181
McDonald, Matt, 151, 152
Militarisation, 18, 35, 158
Military-centric securitisation, 6
Mining residues, 186
Ministry of Communications and Information Technology (MoCIT), 169
Ministry of Environmental Protection, 187
Ministry of External Affairs (MEA), 149–152, 154, 157
Mishmi, 89
  Mishmi mythology, 89
Misinformation, 172
Mission, 75
  rescue mission, 89
Modi, Narendra, 32, 149, 151
Mohenjo-daro, 88
Monetisation, 54
Monetised, 57
Mongolian, 119
Morality, 13
Motor Vehicle Department (MVD), 210

Multilateral cooperation, 68
Multilateral diplomacy, 106
Murder of democracy, 173
Musk deer, 123
Myanmar, 88

N
Naga, 88, 89
Nagaland, 88
Namami Gange, 150
National, 7–10, 14, 15, 30, 32–34,
    39–41, 61, 63, 65, 66, 68–75,
    86, 88, 99, 103, 107, 108, 117,
    119, 124, 128, 129, 152, 153,
    156, 158, 183, 192, 203, 205,
    209, 212
National Action Plan on Climate
    Change (NAPCC), 69, 75, 210
National boundaries, 66
National Development and Reform
    Commission (NDRC), 101–103
National Disaster Act (NDA), 48, 149
National identity, 73
National Institute of Hydrology
    (NIH), 50
National interest, 54, 100, 107
National Party Congress
    17th National Party Congress, 180,
    196
National policies, 4
National security, 34, 98, 104–107,
    146–149
National Water Development
    Authority (NWDA), 50
National Working Conference on the
    Environmental Protection, 181
Naturalisation, 52, 56
Natural resources, 12, 99, 105, 116,
    117
Navratre, 88
Neoliberal, 49, 50, 53, 54
Neoliberal capitalism, 55

Nepal, 171
Net zero, 202, 209, 210
New Delhi, 31, 32, 39–43, 134
New Energy Automobile Industry
    Development Plan, 180, 196
*New Energy Vehicle Industrial
    Development Plan*, 191
New Energy Vehicles (NEVs), 180,
    181, 188–192, 196, 197
    NEV policy, 181, 190
New International Economic Order
    (NIEO), 137
Nickel, 207
NITI Aayog, 54
Nizhneleninskoe, 120
Non-governmental organisations
    (NGOs), 10, 11, 16, 37, 67, 98,
    146, 164, 170
Non-normative, 18, 36
Non-proliferation, 65
Non-traditional security, 33, 105
Norm, 37, 41, 60–63, 65, 73, 88,
    106
Normative, 5, 18, 34, 36, 37, 64, 86,
    119
Norm entrepreneurship, 65
North Atlantic Treaty Organisation
    (NATO), 130, 131
Northeast India/North-East India, 88

O
One Belt, One Road, 117
Onuf, Nicholas, 60, 63

P
Panchamrita, 151
Panchayats (Extension to Scheduled
    Areas) Act, 54
Panel fixed effect model, 190
Pan-securitisation, 129–132,
    134–136, 138

Parekh, Bhikhu, 83, 84
Paris Agreement, 39, 42, 67–69, 75, 104, 210
Paris Conference 2015, 42
Participation, 8, 9, 14, 15, 19, 44, 54, 62, 63, 71–73, 75, 76, 84, 99, 135, 136, 154, 166–168, 173, 180, 187
Particulate matter, 202
Partnership/partnerships, 70, 74, 116, 189
Party-state, 11, 13, 15, 17
Peksen, Murat, 190
People-centred approach, 154
People's Republic of China (PRC), 11, 13, 15, 40, 98–109, 120, 123, 165, 186
Per capita, 38, 40
Pesticides, 186
Petrochemical plant, 120
Plagemann, Johannes, 170
Poaching, 92, 123
Policy/policies, 4–7, 9, 11–18, 31, 35–37, 40, 41, 43, 44, 48–50, 52, 55, 56, 60–67, 69–76, 84, 99–101, 103, 104, 107, 109, 117, 146, 149, 150, 152–158, 162, 168–171, 174, 180, 181, 184, 186–188, 190, 191, 197, 202, 204
Policy framework, 73
"1+N" policy framework, 154
Policy intervention, 7, 8, 19, 187, 188
Political agenda, 98, 107
Political choice, 129
Political processes, 7, 129, 130
Political security, 105–107, 157
Pollutants, 118, 122, 202, 207
Pollution
    air pollution, 41, 104, 186, 187, 191
    marine pollution, 187
    soil pollution, 73, 186
    transboundary pollution, 120
Prayagraj, 150
Predator, 90
Presentism, 86
Primorsky, 116
Privacy, 171, 174
Profit maximisation, 188
Pro Planet People, 151
Propulsion technologies, 191
Proton-exchange Membrane Fuel Cell (PEMFC), 191
Psycho-sociality, 49
Public health, 65, 183, 188
Public opinion, 8, 169
Public participation, 8, 9, 11, 15–17, 19, 168, 169, 180, 188
Public Private Partnerships (PPPs), 51
Puli Kali, 88

Q
Quality standards, 186

R
Radioactive, 118
Rare earth metals, 207, 208
Realpolitik, 101
Red Book crane, 124
Referent object, 6–8, 30, 33, 35–37, 44, 106, 107, 128, 147–153, 155, 156
Reliance Industries, 52
Renewable energy, 9, 10, 13, 17, 41, 42, 63, 70, 72, 74, 102, 149
Republic of Buryatia, 122
Research and development (R&D), 70, 189–191, 211
Resilience, 48, 108, 155–157, 171
Resilience strategy, 108
Resources

living water resources, 119, 121
natural resources, 7, 12, 14, 71, 99, 105, 116, 117, 158, 204
Responding to Climate Change: China's Policies and Actions (RCC), 152, 153, 157
Responsibility to Protect (R2P), 132
Responsible great power, 100, 103, 104, 106, 108
Rockefeller foundation, 48
Royal Bengal Tiger, 90
Russia
Russian, 116, 117, 122–124, 207
Russian-Chinese Agreement, 121
Russian Federation, 117–129, 123, 124
Russia-Ukraine war, 207

S
Sahu, Anjan K., 10, 18, 30, 31, 33, 36, 37, 101
Sando Bawdiani Dukhu, 90
Saran, Shyam, 32
Scallop, 123
Scientific forestry, 92
SCO, 154
Sea-rises, 48, 49
Second Industrial Revolution, 205
Securitisation, 5–9, 12–14, 16–19, 30, 36, 44, 49–53, 55, 56
Securitisation theory, 5, 49, 57
Securitising actor, 7–9, 19, 31
Separatism, 106, 107
Shandong, 185
Shengtai wenming, 13, 180
Shero-vali-Maata, 88
Sitaraman, Nirmala, 205
Small Island Developing States (SIDS), 138
Smart City Mission (SCM), 51
Social constructivist, 5, 6, 60, 151

Socialism with Chinese characteristics, 101, 105–107
Socially driven collaborative governance model, 166
Social media, 11, 162–174
social media platforms, 162, 163, 167, 169–174
Social stability, 8, 104, 188, 203
Solar energy, 70, 72, 75
Solid Oxide Fuel Cell (SOFC), 191
Songhua River, 120, 121
Sovereignty, 68, 99, 100, 106, 108, 129, 132, 135, 157, 170, 172
Special Economic Zones (SEZs), 18
Special purpose vehicles (SPVs), 51
Speech act, 7, 9, 128, 137, 138, 150
Stakeholders, 8, 16, 55, 66, 67, 163, 164, 166
State-centrism, 82, 86
State council, 191
State Environmental Protection Administration (SEPA), 183, 187
State-Owned Enterprises (SOEs), 17, 189
State policy, 118, 187
Strategic, 73
strategic alliances, 157
strategic cooperation, 116
Strategy for the Environmental Safety, 117
Sustainability, 15, 70, 71, 73–75, 117, 156, 180, 181, 183, 188, 197, 205, 208
Sustainable development, 13, 63, 69, 71–76, 118, 136, 152, 154, 169, 180, 183, 187, 188, 196, 204, 209
Sustainable Development Goals (SDGs), 73, 151
Swachh Bharat Abhiyan (SBA), 51, 150
Swatch Bharat Abhiyan

Clean India Mission, 75

T

Technology/technologies, 68–70, 72, 74
    emission-reducing technologies, 101
    technological progress, 189, 191
Technology Transfer South–South Cooperation Center, 136
Techno-managerial, 53
Terrorism, 35, 105–107, 149
Theoretical paradigm, 92
Three-phase policy, 190
Tiger, 88–91, 123
Transboundary, 19, 117–121, 124
Transboundary zones, 120, 122
Transnational, 62, 64–66, 74, 118, 149
Transnational advocacy networks, 66
Transparency, 42, 50, 118, 162, 166, 173
Transport sector, 202, 203, 205, 206, 210, 211
Tree frog, 123
Trepang, 123
Tsunami, 56, 156

U

UAPA, 170
UN 2030 Agenda for Sustainable Development, 153
Underdevelopment, 34, 86, 133
United Kingdom (UK), 30, 129
United Nations Conference on the Human Environment, 187
United Nations Environment Programme (UNEP), 102
United Nations Framework Convention on Climate Change
(UNFCCC), 13, 37–39, 98, 133, 138, 202
United Nations Security Council (UNSC), 30, 35, 98, 99, 107, 109, 129–136
United States of America (USA)/ United States (US), 30–32, 36, 38–44, 52, 53, 103, 104, 129, 135, 136
Universalism/universalisms, 86, 98, 107, 108
UN Peacekeeping Operations (UNPKOs), 153
Uprising
    1857 Uprising, 92
Urban, 48, 51–54, 75, 167
Urbanisation, 48
Urban Local Bodies (ULB), 51
Urban population, 162, 173
US Military Advisory Board, 148
Ussuri, 121

V

Vehicle and Vessel Tax (VVT), 190
Victorian, 90, 91
Virile imperialists, 90

W

Wæver, Ole, 98, 102, 107, 129, 137, 146–148, 150
Washington, 31, 40–43, 136
Waste management, 73, 75, 187
Watercourses, 119
Web of causality, 129
Wendt, Alexander, 6, 60–62
Western Ghats, 6, 52
Western political theory, 87
Westphalian, 87
Wildlife Conservation Society (WCS), 89
World Bank, 53

World Economic Outlook (WEO),
    184
World Health Organization (WHO),
    186
World Meteorological Organization
    (WMO), 67
World Trade Organization (WTO),
    12, 189
World Wildlife Fund (WWF), 121

X
Xiaoping, Deng, 189

Xu, Chenggang, 188

Y
Yangtze, 108
Yule, George, 90

Z
Zemin, Jiang, 100
Zhejiang, 185